The Power of Light

The Power of Light

The Epic Story of Man's Quest to Harness the Sun

Frank T. Kryza

McGraw-Hill
New York Chicago San Francisco
Lisbon London Madrid Mexico City Milan
New Delhi San Juan Seoul Singapore
Sydney Toronto

The McGraw-Hill Companies

1 2 3 4 5 6 7 8 9 0 DOC/DOC 0 9 8 7 6 5 4 3

ISBN 0-07-140021-4

McGraw-Hill books are available at special discounts to use as premiums and sales promotions, or for use in corporate training programs. For more information, please write to the Director of Special Sales, Professional Publishing, McGraw-Hill, Two Penn Plaza, New York, NY 10121-2298. Or contact your local bookstore.

This book is printed on recycled, acid-free paper containing a minimum of 50% recycled de-inked fiber.

This book is for
Natalie and Nicholas

Contents

Introduction

The late industrial revolution was a time of intense technological excitement in America, an era of heightened possibilities and palpable progress. In the half-century that followed the Civil War, work began on the Brooklyn Bridge and other great spans across the nation's rivers, while steam-driven trains rolled along the elevated railways of American cities. Urban streets were newly lit by electricity, while telegraph and telephone increasingly linked the whole nation. It was a time of wonders like the phonograph, the radio, the automobile, roll-film cameras, aerial flight, and moving pictures—all proliferating with such rapidity that the public was prepared to believe in anything. It was a time to be amazed.

"My God," cried the emperor of Brazil when he heard a courtier's voice emanating from the handset of a telephone at the Chicago World's Fair, "it *speaks Portuguese*!" It has been well said (by Sir Arthur C. Clarke, among others) that any radically new technology will appear to the uninitiated, even if they are well-educated, indistinguishable from magic. And what response can human beings bring to magic but a sense of wonder and awe?

This was a moment in history when anything seemed possible, and when many things that turned out not to be possible were nonetheless broached by serious and prudent men. It was a time for conjoining oceans with monumental canals. It was a time for launching flying machines (though none would be successful until the dawn of the new century). Plans for submarine vessels were mooted and many were built. And it was a time to find new sources of energy to power the new age. Why not harness the energy of the sun? If coal could make steam in a boiler, why couldn't concentrated sunlight do a better, cheaper, cleaner job? It seemed an attractive proposition. Many pursued it. And yet, the story of solar power in the steam age has remained largely untold.

When we think of harnessing solar power to do useful work, most of us imagine silicon cells, not steam engines. We think of the twentieth and twenty-first centuries, not the nineteenth. Many assume the first serious solar power initiatives in the United States took root a few decades ago during an "energy crisis," while angry citizens first waited in gasoline lines. Held hostage by an oil embargo, America seemed briefly receptive to warnings that we faced economic pain and political upheaval unless we buffered our reliance on petroleum (much less of it imported in those days than today). That crisis soon passed. By the early years of the Reagan administration, government incentives for solar power had largely disappeared (along with the solar panels Jimmy Carter had installed on the roof of the White House).

In fact, modern efforts to design and build mechanical devices to convert solar energy into more useful forms began a century earlier, at the height of an industrial revolution that was itself founded on the premise of inexhaustible supplies of coal. Those nineteenth-century solar machines, many of which you will read about in these pages, were built, in turn, on a scaffolding of solar investigation and experiment that stretches back into the dim gloaming of history.

The Power of Light is about men attempting to harness the energy of the sun to do work, but it differs from other books about solar power in that it ends where most begin—at the cusp of the nineteenth and twentieth centuries. It tells the story of the nineteenth-century American inventor Frank Shuman, who went to Egypt to build a gigantic solar-powered steam engine on the banks of the Nile. With British backing, Shuman wanted to use solar energy (rather than imported coal) to fuel pumps that would power vast irrigation systems in the Egyptian desert. These would take Nile River water to fertile tracts of land previously thought too high above the river valley to benefit from the Nile's bounty. Shuman succeeded, but his was a brief epiphany in the story of man's determination to tap sunlight as a practical source of mechanical power. With the advent of the First World War and a nascent petroleum economy, his successes were soon eclipsed—and almost forgotten.

The Power of Light also tells the stories of other early solar power visionaries: Archimedes, who was long believed to have battled Roman sailors on the island of Sicily by setting fire to their ships with mirrors; Leonardo da Vinci, who was likely the first scientist to suggest that solar power be used for commercial purposes; Augustin Mouchot, who built the first large-scale solar steam engines in Europe; William Adams, who tried to bring solar technology to British India; Aubrey Eneas, who built solar machines to irrigate the Arizona desert; and a host of others.

For the most part, writers who focus on the science and technology of the nineteenth and early twentieth centuries like to recount the great successes of that era—the triumphs of Diesel, Marconi, and Edison, for example. We read of Thomas Edison laboring over the invention of the lightbulb, bringing dozens of different filaments to incandescence with an electrical current and failing for months to find a durable vehicle for his new invention—until, at last, he develops a source of illumination that has persisted largely unchanged to this day. The story has a mythic quality and has inspired generations of schoolchildren. Stories like that need to be told, for they help us understand who we are and where we come from.

And yet, as a reader of nonfiction who has sought entertainment as much as knowledge from books, I have always preferred the quiet byways and temporary dead ends of science and technology, for that is where the untold stories lie. Because they are fresh, these accounts have gripped my imagination as no new account of Edison ever could. I hope they will grip yours as you read this book. The exploits of lesser-known scientists and engineers of the nineteenth century have always provided a bounty of unusual tales. Among their stories one finds a kind of furious technological quest equal to any fictional adventure of an H. G. Wells or Jules Verne hero. This is so because the late industrial revolution was a time of tremendous physical innovation spurred by an energetic if naive arrogance. By the nineteenth century, many students of technology had begun to foster the previously heretical view that nature should be studied not so much to be understood as to be *mastered*—to be harnessed, controlled, and dominated by human beings. It was an epoch of *applied* science, especially in America. The development of these applications lend themselves to some wonderful historical narrative.

In the pages that follow, I have alternated chapters that recount Frank Shuman's efforts to build large solar machines in Philadelphia and Cairo with sketches of earlier episodes in the history of solar invention, some going back to classical times. I have done this in an effort to provide an historical context for Shuman's achievements and to show that the inventive directions Shuman (and some of his close contemporaries) took in the late nineteenth century were dictated in part—perhaps largely—by their having studied a history of solar invention rich in precedent, a history better known to students of science in those days than today.

Perhaps modern readers will view solar-powered steam engines as having sprung up in a kind of intellectual vacuum, to see them as creations that are, so to speak, so "over the top," so outside the mainstream of the science and technology even of the nineteenth century that they risk being viewed as eccentric (though fascinating) irrelevancies. By relating stories of earlier solar exploits, I hope to make some of the connections, to show the links with the science and technology of the industrial revolution, demonstrating that these solar devices were not irrelevant then and may have more than historical relevance now.

The other reason—perhaps the more compelling one—that I included these stories is because I hope readers will find them as entertaining as I did.

I bring no political agenda to this book. I wrote it not from a belief that solar power's time has come, nor out of a zealot's need to chastise or proselytize. Neither economics nor psychology suggests to me that there is reason to be sanguine about the rapid growth of solar power in the short term. On the contrary, all the evidence suggests that the era of hydrocarbon fuels is far from over.

In the longer term, who can say what the future of solar power will be? Harnessing the sun's energy is one of humankind's oldest fantasies, ranking with perpetual motion and the transmutation of base metals into gold. If in years past it has proven as elusive as those other dreams, that seems to be no longer true today and may prove less so in the future.

A scientific discovery is as much a beginning as an end, and solar-generated steam has found new uses in recent years. Visitors to the Mojave desert outside Barstow, California, cannot help but notice a 300-foot tower surrounded by concentric circles of nearly 2000 giant mirrors, called heliostats. On sunny days in the 1990s, those mirrors reflected the concentrated heat of the equivalent of 600 suns onto black absorption panels at the top of the tower, heating molten salt contained in the pipes within to 1000 degrees Fahrenheit. The salt was pumped into a heat exchanger near the foot of the tower to make high-pressure steam, which in turn powered a turbine generator capable of producing 10 megawatts of power, enough to serve 6000 homes on the Southern California Edison utility grid.

Though the Barstow demonstration plant, called Solar Two, was mothballed at the turn of the millennium (only months before California's power crisis), similar plants may soon compete with conventionally produced power to attract institutional buyers, at least in certain parts of the world. Solar Two and similar modern installations are projects that nineteenth-century solar engineers would surely be proud to call the progeny of their own inventions. Their intellectual and practical lineage is clear, even across the span of a hundred years.

The inventions and discoveries of every age provide a platform for those of succeeding eras. I have no doubt that new chapters in the story of man's determination to use solar energy as a substitute for fuel remain to be written. If those efforts succeed on a scale that past efforts have not, the time may come when the history of solar power in the nineteenth century will become as familiar as the exploits of Edison are today, for they will then have become part of the great mainstream of technology, not merely a curious tributary that (perhaps temporarily) ran dry.

Frank Kryza
Dallas, Texas
July 4, 2002

The Power of Light

1
Philadelphia's Solar Wizard

In October 1911, a stout, well-dressed man with a reddish Teddy Roosevelt mustache, pince-nez, and a black bow tie prepared to board the Cunard Steam-Ship Company's SS *Mauretania,* then docked at Pier 94 above Lower Manhattan's North River, her four black funnels releasing tapers of steam into the crisp Atlantic breeze. The *Mauretania,* then the fastest transatlantic liner, holder of the coveted "Blue Ribbon" for the speediest crossing ever from Great Britain to the United States, was queen of Cunard's fleet. The man was Frank Shuman, on his way to Cairo via London with plans to install the largest solar-powered steam engine ever built, a 1000-horsepower behemoth that would pump millions of gallons of Nile River water onto parched Egyptian cotton fields every day.

Shuman presented his ticket and passport to a customs official who stamped his travel documents, signed them, and waved him up the covered gangplank, a canvas-shrouded tunnel through which he reached the ship's foyer. There a white-jacketed steward greeted him and escorted him to his well-appointed suite.

Shuman was eager for the ship to get underway. He had picked the *Mauretania* especially for her comfort and speed; it was well advertised that she could race across the Atlantic at a record 26 knots, making the crossing to England in just over five days. Smaller than her sister ship *Lusitania,* the four-screw liner consumed 1000 tons of coal a day and required a "black gang" of 324 firemen and trimmers to feed her. Shuman, an engineer who had designed many an engine and turbine himself, soon got the *Mauretania*'s second mate to give him a tour of the vessel's 78,000-horsepower Wallsend Slipway steam turbines, among the first turbines ever to be installed in a passenger ship. He astonished his guide by quickly calculating that the mechanical power generated by these machines in a single crossing of the Atlantic was enough to lift and put into place every stone of the great pyramid at Giza. Such was the industrial might at human beings' disposal in the age of steam: the human race, at last, had learned to *manufacture power*, a feat accomplished in no other epoch.

Though he is little remembered today, Frank Shuman by 1911 was a celebrated inventor. He had hundreds of patents to his name, including one for "wire glass," a reinforced glass for skylights and windows in railroad stations and other large architectural spaces, an invention hailed across the country and one that made him independently wealthy before he was thirty. He developed a widely used process for making concrete piles in riverbeds and coastlines, another money-maker. He would go on to invent "Safetee-Glass" by welding together two sheets of plate glass with "pyralin," a clear epoxylike goo, for automobile windshields, goggles, and machine tool guards. In an era when splintering windshields were an often fatal hazard of driving, Safetee-Glass was an immediate commercial success, and a lifesaver.

Shuman also pursued less profitable inventions that were closer to his heart. The year before his trip on the *Mauretania*, he had developed a low-pressure steam engine. This device, for which Cornell University would award him an honorary master's degree in 1916, was considered revolutionary: The action of the intake and exhaust valves was four times more rapid than in any existing machine, clearance was reduced to one-quarter, and cylinder condensation was reduced to a third of that in ordinary reciprocating (back-and-forth) steam engines. To finance and market this device, he had incorporated Shuman Engine Syndicate, Ltd., in London. And now an even bolder venture was the object of his Atlantic crossing.

Frank Shuman was already a successful and wealthy man. He was also a man on a mission. His low-pressure steam engine was developed with a new source of energy in mind, to replace the coal then universally used to power steam engines. That new source of power was the sun. To finance the development of practical solar power, Shuman's British partners had formed the Sun Power Company in 1910. The firm chose Egypt—then still a British protectorate, where land and labor were cheap and sunlight filled desert skies—to build their demonstration project: a massive solar-powered steam plant that would take Shuman's ideas from the drawing board to reality, and onto the pages of the world's newspapers.

As the *Mauretania* sailed out of New York Harbor, Shuman wasted no time settling into the comfortable routine of the floating palace. The Atlantic passenger trade by 1911 had evolved as far as it ever would from what Charles Dickens, decades earlier, had scorned as "travelling on a hearse with windows." Before the *Titanic* disaster (in 1912) and the sinking of the *Lusitania* (in 1915) changed perceptions irreversibly, ships like the elegant and spacious *Mauretania* were thought to combine absolute safety and convenience with fairy-tale opulence at sea. The print shop issued a daily newspaper, including news items received by wireless. The library boasted 10,000 books and a massive wood-burning fireplace, creating the ambiance even in midocean (as soon as one got used to the gentle rolling of the ship) of a Mayfair or Philadelphia club.

It was to the *Mauretania*'s library that Shuman repaired on that first morning of his voyage, installing himself at one of the three mahogany partners desks available there. He would spend a good deal of time writing in the next five days, for he wanted to hone the marketing pitch he would make to British backers at the round of dinners and scientific society meetings his consulting engineer, A. S. E. Ackermann, had arranged for him in England. At these he hoped to raise subscriptions to the equity of the Sun Power Company (Eastern Hemisphere) Ltd., substantially beyond the hundred thousand dollars he had already collected in the United States. In the quiet of the ship's library, Shuman organized the ideas he hoped to present to wealthy backers:

> *You will at once admit that any businessman who was approached several years ago with a view to purchasing stock in a flying machine company would have feared for the sanity of*

the proposer. And yet, after it has been shown conclusively that it can be done, there is now no difficulty in securing all the money that is wanted, and rapid progress in aviation is from now on assured.

We will have to go through this same course with solar power and I am confident we will achieve the same success.

Shuman had already demonstrated to his own satisfaction that solar power could be put on a commercial footing, at least in certain parts of the world such as Egypt's Nile Delta. The science was not complex. It never was with Shuman. Nor was the economics. He always reduced things to first principles.

Frank Shuman, like Edison and other American inventors of the late nineteenth century, was self-made and, beyond high school, largely self-educated. Born in Brooklyn on January 23, 1862, he was the son of first-generation German-Americans, Armin and Elizabeth Seestedt Shuman. His grandfather, Carl Frederick Schumann (the family later dropped the "c" and second "n") emigrated from Schleswig Holstein's Elbe River Valley to settle at Trenton, New Jersey, in 1848, founding what would become a large family. Frank's father, Armin, served in the Federal army in the Civil War and was later rewarded with a civil service clerkship in Charlottesville, Virginia.

Young Frank attended Charlottesville public schools for three years, after which his parents encouraged him to develop his abilities on his own, without the benefit of teachers. His extraordinary curiosity about the physical world spurred him to become one of a new breed of autodidact inventors, more common in that era than today, who valued above all the ability to convert scientific theory to practical use.

Despite his lack of formal education, at 18 he landed a job as a chemist with the Victor G. Bloede Company, a large manufacturer of aniline dyes based in Parkersburg, West Virginia, a coal mining hamlet nestled in a dogleg of the Ohio River northwest of the Appalachian foothills. His work with Bloede took him on business to the nearest big city, Philadelphia, 360 rail miles away. There he often

A rare photograph of Frank Shuman, with his autograph. This portrait was used to illustrate his entry in the Cyclopedia of American Biography in 1926. It appears to have been taken about the time he returned from Egypt at the outbreak of the Great War.

stopped to see his uncle, Francis Shuman, president of the Tacony Iron and Metal Company, which had cast Alexander Milne Calder's 37-foot-high, 28-ton bronze statue of William Penn for the newly built Philadelphia City Hall, the brooding, crenelated Victorian castle that stands to this day at Center Square at Broad and Market Streets.

The William Penn bronze was to be the tallest statue on any building in the world. In 1891, the Tacony works began manufacturing the huge iron plates that were to brace the statue's 547-foot supporting tower. Appalled at the estimated annual maintenance cost of $10,000 to scrape and paint the cast-iron plates, the City Hall architect asked the Tacony firm to come up with a maintenance-free surface. Given the visibility of the Penn tower project, Francis Shuman invited a half-dozen prominent industrial technologists to ponder the problem. Meanwhile, Frank Shuman began tinkering with plating machinery in his uncle's lab during weekend visits.

Using small sample iron plates in Tacony, the Shumans and New York engineer John D. Darling, who had pioneered a method of electroplating aluminum, developed a new plating process using a thick coat of pure copper covered by a thin film of aluminum, creating a weatherproof seal that would require no maintenance. The use of the wonderful lightweight new metal on such a large scale attracted the attention of the editors of *Scientific American*, who published a report in October 1892 about the Tacony experiments using aluminum in its newfound industrial application. The editors lauded the "indestructible qualities and high resistance to corrosion" of aluminum.

Frank Shuman's work on the Penn statue tower brought him into the inner circle of Darling and other well-known (and well-heeled) East Coast inventors, giving him the connections he needed to finance development of an invention he had been working on in secret. This was the project that would soon make him a wealthy man, freeing him to pursue his own dreams. He quit his job at Victor Bloede and left for Tacony, a Philadelphia suburb, where he moved in with his uncle Francis who took him on at the iron works as an extra hand. At night, he worked alone on his invention.

The late nineteenth century was a time of monumental public architecture in America—soaring public buildings, cathedral-like train stations, and post offices modeled on Greek temples. In the days before electric lighting, these buildings usually incorporated vast skylights to admit daylight, the architectural culture of the day placing great value on natural light. Plate glass, the manufacture of which had been pioneered in neighboring Pittsburgh, had become a cheap commodity, though it was a fragile and dangerous building material. Under the weight of snow or the stress caused by the vibrations of passing trains, skylights often shattered, raining sudden death upon or maiming the unfortunates standing beneath. Newspapers of the 1880s and 1890s are full of stories of collapsing skylights in public buildings. The standard corrective was to hang thick circus netting under the skylight to catch falling glass before it could wreak mayhem at ground level, though this practice was only partly successful. Sharp, spearlike fragments often slipped through. Moreover, the nets themselves were unsightly, defeating the aesthetic intent of skylights.

Frank Shuman's new invention made these ugly nets obsolete. Also it was hailproof, endured easily the vibrations of any train, and could support the weight of thick rafts of snow in the winter. It cut almost no light transmission. Once introduced, it developed the reputation of saving lives, making its young inventor something of a popular hero. It also made Frank Shuman's fortune.

The new invention was wire glass, manufactured by a system Shuman patented for forcing ultrathin chicken wire into a molten matrix of plate glass. With backing from Philadelphia millionaires W. L. Elkins Jr. and Jacob Disston (who both were to profit handsomely from their investment and later contributed generously to Shuman's solar projects), Shuman incorporated the American Wire Glass Manufacturing Company in Tacony in 1892, installing Elkins as president. The new plant, considered high-tech enough to merit a cover story in *Scientific American*, could melt 10 tons of glass a day and produce seamless rectangles of wire glass as large as 3 feet by 8. Wire glass was an unqualified public relations success across America, as well as a commercial triumph. Sales of the product soared in every large U.S. city.

Privately wealthy as the result of this venture, Frank Shuman could now strike out on his own. In 1894 he purchased a 5-acre wooded block of land at the corner of Disston and Ditman Streets in Tacony, transforming it into an "inventor's compound" in the manner of Thomas Edison's rural enclave in Menlo Park, New Jersey. It included a well-equipped private laboratory; one of the first telephone systems in Philadelphia, linking all the buildings on the property with one another as well as the outside world; a baronial residence for himself; and smaller houses for mother, brothers, and family members. Shuman was barely 30 years old.

The Greenhouse Effect

Shuman's interest in solar energy as a substitute for fossil fuels originated early in his career, as he later recounted in a memoir he wrote for *Scientific American*. He was pondering possible new uses for glass, a material he continued to regard as having commercial potential beyond the new applications he had already found for it. He knew that among its almost magical qualities, ordinary plate glass is transparent to visible light and long-wavelength ultraviolet radiation. That was why it is used in windows. And yet, glass is simultaneously opaque to the infrared and radiation of longer wavelengths. Most of the energy radiated by the sun is in the shorter wavelengths of the spectrum. This fact, coupled with the unusual property of glass just noted—its selective transparency—is what warms greenhouses.

On sunny days, short-wavelength radiant energy from the sun is transmitted through the glass and enters a greenhouse. The light hits the objects in the greenhouse and warms them. The energy reflected by the warmed plants, clay pots, and rocks is at the longer infrared wavelengths and cannot escape, because glass is as opaque to these rays as a sheet of lead. While the air temperature outside the greenhouse might be low, the temperature inside will soon rise because more energy gets in through the glass than can get out. This odd characteristic of glass, Shuman realized, provides a cheap and simple way to concentrate solar energy without the use of expensive mirrors or lenses.

Shuman was especially interested in the solar experiments of fellow inventor and polymath Reginald Fessenden.[1] Fessenden too had been captivated by the notion of solar radiation as a practical commercial substitute for coal. He reported in the pages of scientific journals how he had blackened a clear-glass thermos flask on the

[1]Fessenden was a radio pioneer credited with making the first true radio broadcast of speech and music (on Christmas Eve, 1906, from Massachusetts). Fessenden had discovered a way to superimpose sound on radio waves and transmit the signal to distant receivers, a technology superior to Marconi's (the dominant radio technology of the time), which could transmit only Morse code.

inside, filled it with tap water, exposed it to direct sunlight on a frigid New England winter's day, and watched as the flask exploded in a haze of steam and glass fragments—the pressure inside rupturing the vessel as the temperature of the water rose above its boiling point. Fessenden cited a dozen experiments in which water had been raised to temperatures above boiling by the sun's rays in this way, without the use of mirrors or lenses. Fessenden later wrote:

> *It was found by experiment that if a vessel were so arranged that the sun's rays could impinge upon it, and if all heat losses by conduction, convection, and radiation were prevented by a theoretically perfect method of insulation, the temperature within the vessel would rise certainly to a thousand degrees Fahrenheit without any effort to concentrate the rays of the sun [beyond the use of glass itself].*

Though he was quick to dismiss "theoretically perfect" conditions as the chimera they always are, Shuman thought he could learn something from Fessenden's work. It provided the germ of an idea. Conscious of the need to keep costs down in developing any new commercial process, Shuman began experimenting with simple "hot boxes" at his residential compound at Tacony. These boxes were cheaply knocked together by assistants in his carpentry shop: flat wood-frame boxes 3 or 4 inches high and 2 to 3 feet square with a black, watertight interior covered by plate glass. He demonstrated with this apparatus that he could raise the temperature of water to just below the boiling point on sunny days in Philadelphia, even with snow on the ground, simply by insulating the box with shredded newspapers and covering it with a single pane of plate glass.

Satisfied that hot boxes could trap solar heat efficiently and at low cost, Shuman placed a miniature boiler in the box, using ether as the "working fluid." He connected the boiler to a $2 toy steam engine, a popular Christmas present at the time. (He borrowed the engine from his youngest son, John Edward, who was accustomed to operating it by burning sugar cubes soaked in lighter fluid in the firebox.) Modified by Shuman to run on ether, the tiny engine chugged away merrily when the sun was shining, demonstrating in principle—if only at the level of a workbench curiosity—that solar energy could be transformed into mechanical power.

Shuman increased the size of the boiler to a 2-inch steam pipe 16 feet long, enclosed now in a series of wooden hot boxes linked by pipes. He used this apparatus to calculate the amount of energy the system absorbed and could make available for mechanical conversion.

To take his experimentation further toward the practical, Shuman expanded the field of hot boxes to an area 60 feet by 18 feet, with the larger pipes immersed in the solar-heated water as before, to create a collection area of 1080 square feet. The pipes again contained ether as the working fluid in a sealed circuit. Using this larger boiler, Shuman was able to run a modified full-scale steam engine to produce 4 horsepower. He connected the engine to a water pump.

To gauge public reaction to solar power in a practical form, Shuman advertised his invention in handbills he printed up in the late summer of 1907. Distributed in Philadelphia and its suburbs, these invited anyone who was interested to come to the Disston Street compound to witness in person "an exhibition run of the first practical solar engine—on any clear afternoon between 12 and 3 PM." Shuman was proud of this contraption. It awakened in him the salesperson's itch to promote it, for the pitchman in him was never far from the surface.

He wrote personal letters to influential Philadelphians to promote the new machine. In a letter dated August 13, 1907, he told Judge Thomas W. South, a financial backer:

> *We have the Sun Engine in good running condition now. I am about ready now to give a public demonstration of this, and want to get some newspaper reporters present so it may get into the Sunday issues.*
>
> *This is certainly worthy of all interest as it is the start of a new era in mechanics; and for our purpose, we want to make as much publicity of it as we possibly can. . . . I want a big splurge.*

Shuman's first Tacony solar plant. Note the water gushing out of the pipe at right.

> *From what we have got out of this engine, I am sure it will be an entire success in all dry tropical countries. It would be a success here on any sunshiny day; but you know how the weather has been. We are going to run this engine about 11 o'clock today, and possibly this afternoon; and I would like to have you get as many of our influential friends as you know to come up and see it. We stand ready to run it any day the sun shines at all bright.*

In return, Shuman asked that visitors "acknowledge receipt and say when you will come."

They came, and in droves. Shuman extended the exhibition well into the fall and winter. The demonstrations continued to attract crowds well into 1909. A Shuman relative, then a 12-year-old, later recalled the stocky inventor, cane, hat, and bow tie in place, escorting visitors around his compound to show them the solar machine. Children, especially, loved the tours.

Improving the Sun Machine

Though visitors may have found him apparently satisfied with his prototype, Shuman was already having second thoughts about its basic design. The problem was the ether. Ether has a low boiling point, making it conducive to use in "hot box" solar collectors, which never generate the high temperatures of a coal furnace. Shuman knew that one of the main handicaps of high-temperature engines was the prodigal quantities of energy they threw off through radiation, conduction, and convection. In early steam engines, only 2 percent of the energy locked in the coal was ever converted into mechanical energy. Even the more efficient steam engines of Shuman's time threw off as waste heat about half of the energy released from the coal burned to run them. This occurred in part because of the steep gradient between room temperature and the temperature in the poorly insulated firebox—much of the heat in the furnace simply escaped, warming up the air in the building that housed it. This lost heat could do no useful work.

Using various ethers (which boiled at anywhere from 70 to 120 degrees Fahrenheit, depending on the type of ether used and its purity) cut back the heat gradient between the firebox and the surrounding environment sharply, alone reducing heat loss.[2] But this advantage came at a cost. Even at respectable pressures, ether exerted little force to drive a motor because of its low specific gravity. It could never develop the "head" of a powerful steam engine; the ether was just too thin. This limited the work any ether-powered system could do. No working locomotive, for example, could ever be powered by an ether engine.

There was also a less technical objection to ether that was perhaps more important. The accepted source of mechanical power in most applications in 1900 was the *steam* engine. Shuman understood popular psychology well enough to realize that the general public—and certainly his fellow engineers—would take solar power seriously only if he could find a way to generate real steam with his appara-

[2]The "heat gradient" is the difference in temperature between the ambient air and peak temperature in the boiler. The greater this difference, the more heat energy is likely to be lost to the outside entvironment. The principle is easily demonstrated with two cups of coffee, one scalding and one tepid. All things being equal, the scalding coffee will lose heat to the surrounding air more quickly than the tepid cup.

tus, steam to run a conventional steam engine, not one modified for exotic, low-pressure, low-temperature fluids. He had to adapt his solar apparatus so that it could power *any* steam engine, he recognized, if the solar idea was to catch on. This was a challenge that would haunt his work, and one he never really mastered.

First, he insulated the boxes more efficiently, to make hotter water faster. The key to doing that was to cover the hot boxes with a second layer of plate glass, the panes separated by an inch of dead air. This design retained heat better, pushing the water temperature up more quickly.

"Were no steam made in these vessels," he later wrote, "the temperature therein would go up to 350 degrees F. in latitude 40 north [i.e., around Philadelphia], possibly easily to 450 degrees F. near the equator. The production of steam at atmospheric pressure, however, keeps the temperature in the vessels down to 212 degrees F., and whatever excess of heat is produced by the sun's rays over and above that lost is converted into steam and may therefore be utilized" in powering a steam engine. This breakthrough convinced him he was coming closer to solving the problem of using sunlight to make the vast quantities of steam needed to run an ordinary steam engine, and to capture the public's imagination.

Testing Commercial Viability

By 1910, Frank Shuman felt he had grasped the principles of hot-box design well enough to make his first try at a solar-power plant on a commercial scale. Though he recognized that solar power would have the greatest economic value in the tropics, where sunlight is plentiful and coal expensive, he chose to continue his experiments in Philadelphia on the theory that if solar machines worked in cloudy Pennsylvania, they would work even better in a sunnier climate. A full-scale demonstration plant sited at his laboratory compound, he reasoned, would attract the seed financing needed—from rich Philadelphia and New York investors—to take the concept to market.

To his backers, he offered a more mundane reason for staying put in Tacony: "The plant was set up in Philadelphia," he said, "not because it was considered to be a commercial thing there, but because the necessary experimenting with a new plant thousands of miles from home would have been exceedingly expensive."

As he set out now to take his solar experimenting to a new level, he decided to set down a number of what he regarded as axiomatic ideas he had developed about scaling up solar-power plants. In 1911 he spelled these out in the pages of *Scientific American*. Earlier efforts to harness the sun were based on the use of lenses or mirrors to concentrate the sun's rays, or on the use of low-boiling-point fluids, he noted:

> *It has always been attempted to create vapor at high pressure, and then utilize this in the ordinary engine, but with the high temperatures involved, the losses by conduction and convection are so great that the power produced was of no commercial value. Where lenses or mirrors are used, the primary cost of the lenses and the apparatus necessary to continuously present them toward the sun have rendered them impractical. Where fluids of low boiling point, such as ether and liquid ammonia, were used, the results were of little value by reason of the inherent inadequateness of these fluids as power generators.*
>
> *A sun-power plant in order to be practicable, must possess, first, high efficiency; low cost of installation and maintenance; well-marked length of service; and should require no specially trained mechanics for its operation.*
>
> *An ideal plant must be subject to little accident; hence it must lie near the ground in order not to be affected by storms and winds. Each unit [hot box] must be repairable without stopping the operation; construction must be simple and easily understood by the ordinary steam engineer; and wear and tear must be reduced to a minimum.*

Above all, the initial investment required must be kept low, he cautioned, for "this is the rock upon which, thus far, all sun-power propositions have been wrecked."

Shuman was quick to add that it was not necessary that a sun-powered steam engine be as cheap as an ordinary steam engine, for the cost of the fuel in the former, after all, was free whereas coal always had some cost. Back-of-the-envelope calculations revealed that the up-front investment in a solar plant could be as much as double the cost of an ordinary steam-power plant of comparable size and still be profitable. Later Shuman would refine this financial analysis using numbers derived from actual tests:

> *Even if the extra interest is taken into consideration [in financing the solar plant], the fact that after installation no fuel is required is such an enormous advantage as to entirely offset the increased initial cost, and in addition cause great profits.*

Half an Acre of Hot Boxes

This was the operation Shuman was now prepared to build in Tacony, on a scale never before seen. The new plant, constructed in late 1910 and early 1911, consisted of a solar heat "absorber"—a field of 26 banks of hot boxes, 13 on each side of the steam engine, a condenser to return the steam to the hot boxes as water, and a water pump.

The hot boxes and the metal boilers boasted new features. Though housed in the familiar wood-frame exteriors 3 feet square with the double-glazed glass covers he had developed the year before, the clumsy pipes inside had given way to flat, lamellar boilers made of tinned copper painted a dull black. Each of these was also about a yard square, entirely filling the interior of the box. Installed, they resembled large black waffles. The honeycombed metal construction created a boiler less than half an inch thick, increasing the area of water-to-metal contact and more efficiently heating the water.

The hot boxes were also more tightly insulated. The double sheets of plate glass were set loosely into the wooden frames so they could expand with the heat, yet tightly enough to prevent outside air from seeping into the insulating layer of dead air. Between the lower sheet of glass and the top of the boiler was another air space of about an inch, and below the boiler itself, a third air space of about half an inch. Beneath that, Shuman installed a sheet of millboard (waterproofed cardboard) a quarter-inch thick covering a 2-inch layer of regranulated cork, followed finally by a thicker layer of millboard soaked in creosote or linseed oil to protect the box from the elements.

The boxes were mounted on steel A-frame supports that kept them 30 inches above the damp Tacony soil. The frames permitted the hot boxes to pivot so they could be inclined perpendicular to the noonday sun, improving their ability to collect light. Every 3 weeks Shuman's assistants would adjust the tilt to follow the sun's seasonal path across the Philadelphia sky.

Because Shuman was in the glass business and always interested in maximizing use of glass in his inventions, he made one concession to the "lenses and mirrors" crowd. He added a plane mirror a yard square to each side of the hot boxes. These were as cheap as he could

The second Tacony plant. Note the mirrors flanking the hot boxes.

Concentrating solar collectors in the second Tacony plant.

make them—sheets of ordinary Pittsburgh plate glass silvered on one side and waterproofed. The mirrors were held in metal frames set at an angle of 120 degrees to the glazing of the box. The top edges of the mirrors were thus about 6 feet apart while their bases were separated only by the 3-foot width of the collector, effectively doubling the area of sun collection for each hot box from 9 square feet to 18.

The 26 banks of hot boxes each contained 22 hot-box units, for a total of 572 hot boxes in the array. Because each hot box was 9 feet square, they together provided 5148 square feet of primary collection surface, increased by the mirrors to double that, or 10,296 square feet. With gangways separating the rows and space for the engine and pump, the equipment took up half an acre of open ground at Shuman's Tacony compound.

Each hot box was linked to a water supply pipe at one end and to a steam pipe at the other, connected in series. The steam pipes converged into an 8-inch diameter main pipe that carried the steam to the engine. Once the steam had done its work in the cylinders, it was cooled in a condenser and recirculated through the feed water pipes back into the hot boxes. The cycle was watertight and leaks, Shuman reported, were rare.

A New Steam Engine Design

The steam engine itself was a special low-temperature, low-pressure device Shuman had designed and would later patent, a forerunner of the machine for which he would be honored with a degree by Cornell. It was becoming apparent, when he jettisoned ether as a working fluid, that barring the use of expensive concentration devices, solar collectors would never generate the steam pressures of a coal-fired boiler. In these, the flue gases in the firebox reached temperatures of 2500 degrees Fahrenheit or higher, creating an inferno capable of producing pressures in the boiler upwards of 1000 pounds per square inch. Getting an ordinary steam engine to work below these high levels of heat and pressure turned out to be a vexing problem. Throughout 1908, 1909, and 1910, visitors to Shuman's laboratory saw how preoccupied he had become with the challenge. Littering the workroom were model steam engines, many built by Shuman himself, each taking the technology of low pressure one step closer to realization.

The problem was this: In a conventional steam engine, high-pressure steam is forced into a piston-cylinder assembly. As the steam expands to lower pressure, part of the thermal energy is converted into work by pushing the piston forward. The back-and-forth movement of the piston is converted into rotary motion with a crankshaft, which is connected to the machinery to be powered. The expanded steam, now depleted of most of its energy, is sent to a separate apparatus (called a condenser) where the remaining heat and pressure are dissipated, turning the steam back into water. Some of the waste heat from the condenser is used to warm fresh, incoming water that will be used to make more steam. The condensed steam itself is returned to the boiler to be reheated. In many steam engines, substantial amounts of steam spewed out into the surrounding atmosphere through joints and leaks, though it was possible by Shuman's day to create steam engines that were virtually leak-proof. The steam Shuman could make with his solar absorber could not reach the temperatures and pressures needed to make such a device function at normal air pressure.

The engine in the second Tacony plant.

Shuman was aware that since the early 1900s steam turbines had begun to replace steam engines in larger power plants. Turbines were more efficient and more powerful than reciprocating steam engines, with their clunky, friction-causing cylinders and pistons, because a turbine was, in a sense, a much simpler machine. It operated on the principle of the windmill, with the steam causing an internal rotor to spin inside a shaft, with little energy loss. Steam had a huge energy advantage over wind—it expanded in volume with tremendous velocity, as much as 4000 feet per second when it was under pressure. Turbine rotors could harness this power, rotating at speeds that reached thousands of revolutions per minute. This rotary motion was then transferred via gearing to machinery that performed useful work—often a dynamo that generated electricity.

Shuman desperately wanted to adapt a turbine to operate with solar-generated steam, but he had to face the fact that he was using "wet" steam operating at ordinary atmospheric pressure as his source of energy. Turbines, too, work best with superheated steam under high pressure, a fact that handicapped Shuman's design effort. Eventually he decided on a more conventional machine—one he described as "a new type of low-pressure, reciprocating steam engine of great steam economy," a device for which he received a patent years later, in March 1917. By creating a partial vacuum in the steam engine, he was able to convert more of the heat contained in the solar-heated water into mechanical work.

He made the intake and exhaust manifolds in the machine more efficient and reduced water condensation in the cylinders—a consequence of using steam that was not superheated. The main change was the use of larger cylinders, increasing the surface area on which the low-pressure steam could work to drive the pistons. He was pleasantly surprised at the result. His reciprocating steam engine, powered by the hot boxes on a sunny Pennsylvania day, could pump 3000 gallons of water a minute to a height of 33 feet. In 8 hours of continuous operation under good conditions, the plant could produce 4825 pounds of steam, results that compared well with conventional boilers and steam engines of the day.

The Power of Steam

To make it possible to compare different engines and motors, which can vary enormously in size and design, against a single standard, engineers usually rated steam engines in terms of their power, or the rate at which the engine does work over time. This made it possible to compare a low-powered steam engine, which may deliver a given amount of work over a period of hours, with a high-powered motor, which might do the same amount of work in minutes.

By the late industrial revolution, when the science of mechanics had flowered as a subbranch of physics, the units of power of a machine were defined as those of work per unit of time. These included measures such as foot-pounds per minute, joules per second (or watts), and ergs per second. These technical terms had by Shuman's day been largely supplanted, in the popular mind at least, by a term invented by the British engineer and inventor James Watt, the father of the modern steam engine—the unit we still know as the *horsepower*.

When Watt was trying to sell his early steam engines, he found that potential buyers had difficulty understanding just how much work the new-fangled devices could do because technical terms like "foot-pounds" provided no frame of reference for the average businessperson.

Watt, businessperson that he was, shrewdly restated the relative strength of his machines in terms that his buyers could readily grasp:

the horse. His customers were initially mine owners who used horses to haul coal and groundwater out of mines, an expensive proposition that required lots of horses, stables, food for the horses, and men to tend them. Watt wanted to sell the mine owners on the concept of using steam engines to replace horses. He wanted to present them with a back-of-the-envelope cost-benefit analysis that said, in effect, "lease my steam engine, and you will be able to eliminate the care and feeding of 'X' horses from your costs."

To do this, he had to invent a way to express the work done by a steam engine in terms of the equivalent work that could be done by a horse. By testing a number of horses at mine sites, he determined that the average English horse of the time could haul coal at the rate of 22,000 foot-pounds per minute for about 10 hours a day (a foot-pound being that amount of work needed to lift a weight of 1 pound a distance of 1 foot).

In fixing this new unit to state the power of a steam engine in terms of the work done by a horse, Watt arbitrarily increased the calculated figure by one-half to guarantee that his buyers would get full value for their money. Thus was born the "horsepower," a newly defined unit of work equal to a rate of 33,000 foot-pounds of work per minute. Just as Watt predicted it would, this intuitive measure soon came into popular use in the business world.[3] It is still used to compare the relative power of everything from lawnmowers to automobiles to the jet engines on a spacecraft.

Testing the Philadelphia Plant

By lifting 3000 gallons of water per minute 33 feet, Shuman calculated that his solar-powered steam engine was delivering work at a rate of about 25 horsepower. He had designed it to put out 50 horsepower under optimal conditions. To generate 4825 pounds of steam

[3]Thomas Jefferson, on a visit to England to do a little commercial reconnaissance, interviewed Matthew Boulton about the comparative efficiency of his steam engines and arrived at the conclusion that "a peck-and-a-half of coal perform exactly as much work as a horse in one day can perform."

in 8 hours, the "absorber" was putting out something like 80 horsepower. The difference between the 80 horsepower generated by the boiler and the 25 horsepower delivered by the steam engine was energy lost as waste heat.

A good-sized coal-fired boiler in the first decade of the twentieth century—the time Shuman was conducting his experiments—would have been capable of generating 100,000 pounds of steam per hour, consuming 2 or 3 tons of coal to do so, to run a 3000-horsepower steam engine. That would have been a big machine, capable, say, of running an electric dynamo to illuminate a small city. Such a device would have occupied a large plant, with a smokestack to carry away the hot gases from the furnace.

Shuman's first full-scale demonstration project produced an average of 600 pounds of steam per hour, less than 1 percent of what a large coal-fired boiler might produce. He was tapping from sunlight the amount of energy per hour contained in about 30 pounds of coal. Given this was his first try, Shuman thought these results respectable.

In 1910 no less than today, venture capitalists liked to have the results of a new technical process validated by independent outside authorities before committing money to it. To get such an independent assessment of the project, Shuman brought in as a technical consultant the British engineer A. S. E. Ackermann to test the apparatus. Ackermann published his report on Shuman's machine in the journal *Nature* in 1912. Ackermann was enthusiastic:

> *When experimenting in Philadelphia in July, 1910, with a single unit [i.e., a single hot-box assembly] and no mirrors, the maximum temperature I recorded under the lower cover glass was 250 degrees Fahrenheit, and temperatures over 200 degrees Fahrenheit were common. Even in the latter cases steam was formed freely, showing that the temperature of the boiler was at least 212 degrees.*

Ackermann reported that the maximum quantity of steam produced in any one hour was 800 pounds at atmospheric pressure (equivalent to about 100 gallons of water converted into steam):

and while this is by far the greatest quantity ever produced by sun power, it must be pointed out that Philadelphia is by no means an ideal situation for such a plant, for we had to wait weeks to get a cloudless day and then got three in succession. . . .

The plant was built at Philadelphia simply for the convenience of being close to the inventor's house, offices, and laboratory. In places like Egypt, Africa, Arizona, and California, I should expect to get about 25 per cent more steam for the same collecting area.

Shuman's own estimate of the improvement he might see in lower latitudes was even more optimistic—he thought the plant's energy output would rise threefold at sites closer to the equator. Even to his critics—and several were now beginning to emerge—the first test results from Tacony were clearly impressive.

Ackermann's report hinted at Philadelphia's growing smog problem and how it hindered the machine, compounding the losses due to the sun's low angle. Heat gradients caused by cold weather also caused heat loss. In a *Scientific American* essay, Shuman wrote:

It is found by observation that the steam generated in a sun-power plant is reduced largely by humidity, and the presence in the air of smoke, haze, etc. It follows then that the efficiency will be greatly increased when the apparatus is tested in a dry climate free from the atmospheric impurities attendant upon proximity to a large city.

The loss of heat by conduction and convection in northern latitudes is enormous. If the present apparatus is placed in an average air temperature of 100 degrees Fahrenheit, such as obtains throughout all equatorial regions, it is safe to assume that the power will be multiplied three fold.

Shuman was quickly coming to the view that the size of the steam engine that could be supported by solar energy was simply a function of the surface area of the field of hot boxes—the size of the "absorber." He believed that solar-powered steam engines of larger size than the one he had just built were feasible just by scaling up his existing design. Shuman did not think qualitative changes were required. He was already dreaming of much larger machines:

> *In order to be efficient, it is not necessary that the plant generate continuously, inasmuch as the great value of such a plant lies in its use as an irrigation apparatus; it is only necessary that the plant run about eight hours daily. It must, however, consist of units which may be assembled to produce a power plant of any required size, the larger the plant the greater the efficiency.*

He predicted, "It is entirely practicable to produce a sun-power plant in this manner up to 10,000 horsepower and over" though the "absorber" for such an engine would need to be about 300 times greater than the one in Tacony, covering something like 3 million square feet—some 60 acres of hot boxes.

Practical Opportunities

With Ackermann's sunny evaluation in hand, the time had come to market the idea and raise the funds to turn it into another successful business concern. In 1911, Shuman drafted another panegyric to mechanical solar power, this one addressed to his fundraisers. A version of it later appeared in *Scientific American:*

> *Having described the mechanism of the sun-power plant, it remains to discuss the opportunities for its use. [These lay] in those regions in the tropics where the sun practically shines throughout the year, and where fuel is expensive, coal costing in some localities $30 per ton.*
>
> *There is room now for at least half a million horsepower in such tropical fields as the nitrate district of Chile, the borax industry in Death Valley, and for general purposes in places where the outside temperature runs from 110 to 140 degrees Fahrenheit.*

The potential even for modest-size plants that could be built immediately with the technology in his Tacony compound appeared limitless. As an irrigation engine, his solar plants would find useful application around the world—in Egypt, India, Ceylon, the Australian desert, and the deserts of Arizona, Nevada, and California. These were locales where high fuel costs and ample sunshine conspired to make his machine economic under all existing conditions. With characteristic optimism, he wrote: "It may be assumed that 10 per cent of the earth's land surface will eventually depend upon sun power for all mechanical operations."

This was just a question of comparative economics. The biggest incentive for use of solar power, Shuman believed, was the high cost of fossil fuel (mainly coal) in the land areas bounded by the two tropics where sunlight was most plentiful. In Egypt, imported British coal went for $15 to $40 a ton. Shuman thought as early as 1911 that solar steam could compete with coal costs as low as $3 to $4 per ton. That left a wide margin in which he could operate profitably.

Taking the longer view, Shuman also believed there was a far more compelling reason to develop solar power sooner rather than later. He subscribed to the prevailing scientific wisdom at the turn of the twentieth century about the world's fossil fuel reserves—that they were finite, even constrained, and might be depleted within decades, with calamitous results for "civilized humanity." In 1914, in a memorable letter to a scientific journal, Shuman wrote:

> *One thing I feel sure of and that is that the human race must finally utilize direct sun power or revert to barbarism" [because eventually all coal and oil will be used up]. I would recommend all far-sighted engineers and inventors to work in this direction to their own profit, and the eternal welfare of the human race.*

Since coal was first mined in quantity early in the industrial revolution, every generation has had its "fossil fuel Malthusians," and they were particularly vocal at the turn of the twentieth century. Robert H. Thurston, a professor of engineering at Cornell and one of the imminent scientists of his day, captured the conventional wisdom well. Writing in the Smithsonian Institution's annual report for 1901, which (then as now) was printed by the Government Printing

Office and distributed to all members of Congress, he predicted that the probable life of U.S. coal reserves was

> *something like a century before our stock of coal will be so far depleted as to make serious trouble in our whole social system. In Great Britain the case is probably vastly more serious than in the United States, for there the coal beds are far more restricted in area, and already extensively depleted. The same is to be said of Europe.*
>
> *Within a few generations at most, some other energy than that of combustion of fuel must be relied upon to do a fair share of the work of the civilized world.*

Finding Investors

Shuman's writings and speeches reveal him to be, like many nineteenth-century engineers, deeply idealistic, motivated by a belief that he was working for the betterment of humankind. Most of his patents were for safety devices. In developing solar power, he believed he had found a way to benefit countless millions, not merely the users of safety goggles and automobile windshields. But he never allowed his ideals to color his business judgment.

Ever the salesman, he had balance sheets and profit projections on his mind as he sat at one of the expansive desks in the SS *Mauretania*'s library in the fall of 1911, crafting the marketing campaign he would undertake to win over British investors when he made landfall. Solar power would have to be sold as a business proposition that could make money *immediately*, he knew, or it would go nowhere.

He already had a taste of how difficult that task might be. Before leaving Philadelphia, he had gone back to the financial oasis that had funded all his earlier ventures—the "big-money" men of Philadelphia: Disston, Elkins, Drexel, Biddle, Hoopes, and others. But Pennsylvania was America's premier coal country, and these tycoons were already committed to fossil fuels. For the first time, under smoggy skies and in the sunless recesses of the exclusive clubs off

Walnut Street and Rittenhouse Square where he made his proposals to these rich men, Shuman's arguments fell on deaf ears. His former backers reached modestly into their deep pockets, or not at all, to support his new venture, and Shuman realized he would need fresh backing from other sources to make it work. He needed support from people who were not heavily invested in coal and, preferably, who already had large land holdings in the tropics, where solar machines could make a difference to the bottom line right away.

He considered taking his case to California, then on America's growing frontier. He thought Florida, too, held promise. In the end, he set his sights on London, the world's financial capital, a city that was home also to the colonial masters of the globe's sunniest territories.

He could not have imagined, as the harbor pilots guided the huge liner out of the Irish Sea and past the treacherous shifting sand bars of the Mersey Estuary to Liverpool's docks, just how successful he would be there.

Fundraising in England

Practical demonstrations of Frank Shuman's solar technology in the years before World War I would win the enthusiastic support of Lord Kitchener of Khartoum, the British proconsul in Egypt; Sir Reginald Wingate, the iron-fisted ruler of neighboring Sudan; and earn Shuman an invitation from the German Reichstag to accept the equivalent in Deutchmarks of $200,000 in venture capital—a colossal sum equivalent today to millions of dollars—to bring solar power to Germany's growing colonial possessions in Africa.

When he disembarked the *Mauretania* in England in 1911, Shuman planned to spend several weeks—months, if necessary—in London to raise money for his Egyptian venture. He intended to sell his steam engines and their solar collectors to prospective investors on the basis that solar power was cheaper than the fossil fuel alternatives then available in tropical countries. Shuman thought of himself as in the business of marketing cost-effective mechanical power—steam engines that just happened to be powered by the sun.

He did not expect his backers to subscribe to solar power for any reason other than the most self-serving one—that they would profit from it financially. He appeared to have had no environmental or political agenda.

With the internal combustion engine and oil-based fossil fuels still so ascendant at the beginning of the twenty-first century, it may be hard to recall that solar power was considered by some to be, in the two or three decades before the First World War, the energy technology of the future that held the most promise. A few mainstream scientists of that period even speculated that solar power might soon develop into a major world energy contributor. An indication of this is the quantity and kind of coverage solar power technology was getting in the world's science journals. Between 1880 and 1910, there were 48 articles on solar energy as a world energy source in the pages of *Scientific American* alone, then as now one of the popularizers of cutting-edge science. After 1915, when global commitment to crude oil, gasoline, and the internal combustion engine became irrevocable, discussion of solar energy dried up in the pages of science periodicals, falling into an obscurity from which it would emerge only through the development of silicon photovoltaics in the early 1950s.

The commitment that engineers of Frank Shuman's generation made to solar technology can best be understood within the context from which it sprang. With the industrial revolution, Shuman understood that man's place in the world had changed qualitatively: mankind's capacity to do work had previously been limited to his own strength and that of the men and animals he could control—and what power he could extract from wind and water. Shuman told his London backers: "Man's capacity is no longer limited, for he has now learned *to manufacture power* and with the manufacture of power a new era has begun."

New Sources of Energy

Europe's appetite for energy in the decades of the industrial revolution had grown exponentially. The historian of steam power Richard

Hills has calculated that in the year 1780 there was about 85,000 horsepower available to commerce and industry from all sources (other than animals and the wind power generated by sailing vessels) throughout Europe and England. Of this 85,000 horsepower, 10,000 was generated by windmills, 70,000 by waterwheels, and only 5000 by the newly invented Newcomen steam engines, used mainly to pump water seepage out of mines. (To give these figures scale, consider that the Wallsend Slipway steam turbines that propelled the SS *Mauretania* across the Atlantic on Frank Shuman's voyage delivered a little more than 75,000 horsepower at full throttle, so we can say that total commercial power available across Europe and in England in 1780 was not much more than the power generated by a single oceangoing passenger liner 130 years later.)

By 1880, total horsepower available in Europe and England had grown 26 times to 2.3 million, of which steam now constituted 90 percent. By 1911, total horsepower used by commerce and industry reached 10 million, with steam comprising 99 percent. Total commercial energy employed had thus risen by a factor of 115 since 1780. More important, the balance from naturally generated power to manufactured power had shifted decisively and permanently: In 1780, 95 percent of total power used in commercial applications was from natural sources (wind and water), but by 1911, all but 2 percent of power was *manufactured* power—that is, power made from burning coal and harnessing steam.

Cheap coal, the new universal fuel, and cheap iron, the new universal material, replaced wood in its cruder uses. Production of coal went up by 15 times (from 10 to 150 million tons) in the century between 1780 and 1880. In that period, output of iron increased 115 times (from 68,000 to 7,750,000 tons). Both coal and iron were bulky goods needed in enormous quantities. The cost of moving them tied industry down initially to the coal fields and the ore mines, a limitation that needed to be removed if industrial concerns were to be free to situate themselves wherever they chose.

The development of the steam engine had itself been spurred by the pressing need to pump water out of mines and later to replace animals in hauling out coal and metallic ores. Steam engines could do the work of many horses, cheaply. As so often happens when a previously expensive technology suddenly becomes inexpensive, the early availability of manufactured power fed on itself. Because of the growth to which the new manufactured power gave rise, the quantities of materials that needed to be moved around Europe and England themselves increased exponentially.

It was this newfound need to move things around that became midwife to the greatest and most characteristic innovation of the nineteenth century—the rise of the railroads. (Alfred North Whitehead might have demurred. "The greatest invention of the 19th century," he famously observed, "was the invention of the method of invention.")

Tall, soot-encrusted factory chimneys sprang up amid the grimy, gaslit cities of the industrial age, each one proclaiming, in Carlyle's phrase, another "Stygian forge with fire-throat and never-resting sledge-hammers." Iron rails crisscrossed England and Europe like the strands of a spider's web. Manufactured power had come into its own and with it the growth of an industrial economy linked together across Europe. But to manufacture power, you needed new forms of energy in vast quantities. Firewood was scarce; coal was costly and dangerous to mine; petroleum was still an exotic combustible, thought to be rare. What else was available?

It was in this era that men first thought seriously of the sun as a source of energy to fuel the colossal engines of the new economy. To be sure, inquisitive men of earlier times had dabbled with solar energy, but the need for new sources of power had never been so urgent, and those prior efforts were greeted mainly as scientific curiosities. Mastery of the sun seemed tantalizingly close in the age of steam—one had only to substitute the heat of the sun for wood fires and coal flames. The engineers of the nineteenth century worked with forces large enough to give them the sense, for the first time in history, that they were masters of nature, in possession of the instruments they needed to change the conditions of life profoundly. Why not tame the sun's energy?

This was to be the heyday of the first era of solar invention, from the middle of the nineteenth century to the first decades of the twentieth, when Europe and America, in the throes of the late stages of an overwhelming industrial revolution, produced a scientific and technological cornucopia of new inventions. It was a time for the realization of previously unimaginable projects—of human-made canals linking oceans, of railways linking all the nations of Europe and Asia, of balloon flight, of submarines—and a time of speculation about even more fantastic projects that would not be realized.

One of the issues of *Scientific American* Shuman carried with him on the *Mauretania* contained a proposal by a consortium of respected engineers to flood that portion of the Sahara below sea level to create a vast inland sea, transforming North Africa into a watery agricultural paradise. Of course, it was a proposal never carried out, but it *could have* been done, and that was the point. And there were visionaries bold enough to want to undertake it. The ambitions of engineers knew few bounds.

Scores of projects nearly as bold and on a scale just as vast—now little remembered—were mooted in this epoch. Few proposed feats of engineering or invention seemed outrageous or hubristic in the first decade of the twentieth century, surely the last decade when most educated people believed there was *always* a positive relationship between technical progress and the advancement of the human spirit—a notion that would seem dangerously loony after the technological horrors of the Great War. Scientists and engineers would never again be so revered in America and Britain.

The idea that the sun's heat could be harnessed in some way as a source of immense power, Shuman realized, was not a new one, though he thought it was a notion that had been only crudely developed until now. His reading had already taught him that the power of concentrated sunlight was a theme that ran throughout history, going back, indeed, some thousands of years. It was a branch of the history of technology that was endlessly fascinating, a fruitful field of ideas so long forgotten they often seemed new to modern eyes.

2

Shuman's Inspirations: Solar Power in Ancient Greece and in Europe in the Middle Ages

In pursuing his own solar experiments, Frank Shuman had become, early in his career, something of an historian of the development of solar power. He knew that the first instances of harnessing solar energy stretched back into the crepuscular shadows of time, as far back into prehistory as the clay tablet era in Mesopotamia, the Sumerian "land between the two rivers" (the Tigris and Euphrates) in what is modern-day Iraq. Here temple priestesses may have used polished golden bowls as crude parabolic mirrors to ignite altar fires, as they later would in Greece and Rome.

These uses of curved mirrors had likely been discovered independently in China and in Greece. References to "burning mirrors" (the ancient and medieval term for reflectors used to start fires) were found in the earliest written texts, Shuman knew, and some scholars thought it possible that Chinese and Greek knowledge were both derived from a common source in Mesopotamia, India, or Egypt.

These early uses of sun power had always informed Frank Shuman's tinkering. His own suite of solar inventions can be thought of as the apex of a pyramid containing, layer upon layer, two millennia of prior experimentation with the sun and its uses. To understand the path Frank Shuman took and the choices he made, an account of some of these earlier chapters in man's quest to harness the sun is vital.

Kircher's Mirrors

Years before his journey to England on the SS *Mauretania*, Frank Shuman had duplicated the experiments of a medieval alchemist, experiments that changed science's understanding of how the sun's rays could be concentrated. These experiments, as elegant in their simplicity as they were powerful in the lesson they taught about the nature of the sun's heat, could be set up on any sunny day on the grounds of Frank Shuman's spacious compound in Tacony. There, a brick and stone wall separated the garden of his residence from the cleared open space outside his laboratory. It was against this wall that Shuman liked to entertain visitors with a demonstration of "Kircher's mirrors."

In Rome in the 1640s Father Athanasius Kircher had shown that sunlight could be concentrated at a distance with the simplest of tools in sufficient quantity to ignite fires. Kircher, who had fled Germany during the Thirty Years' War, was a Jesuit scholar and alchemist. He spent much of his later life in Rome where, in a book-lined office in the Vatican, he served as a kind of one-man intellectual clearinghouse for the cultural and scientific reports his fellow Jesuits gleaned across Europe and through their far-flung network of missionaries in Asia, Africa, and the Americas.

Kircher, like Roger Bacon before him, was a scholar who revered the written word, but also like Bacon, he believed most of all in the evidence of his own senses. He was one of the world's earliest "natural philosophers," as practitioners of experimental science called themselves before the word *scientist* came into vogue. Personal risk did not daunt him—he had once had assistants lower him in a sling into the crater of Mt. Vesuvius after an eruption in order to study the spewing lava firsthand. In Rome, he assembled one of the first natural history collections. It was housed in a museum that bore his name, the *Museo Kircheriano*.

Father Athanasius Kircher, the medieval alchemist whose experiments with mirrors in Rome proved conclusively that banks of ordinary, flat mirrors arranged to focus the sun's rays on a single spot could produce very high temperatures.

In 1646 Kircher published *The Great Art of Light and Shadow,* a work on optics. Among the novelties he described was a projecting device equipped with a focusing lens and a mirror. It used sunlight as its light source. Kircher also wrote of a primitive method of using artificial light from a candle or lamp for projection of images. Neither projector was practical, but in time others developed it into the "Magic Lantern," a forerunner of modern slide and movie projectors. In what may have been the world's first horror film, Kircher is alleged to have projected a watercolor image of the devil onto the parchment windows of a lapsed parishioner late one night—to marked effect. According to the story, the man speedily resumed church attendance.

Kircher also explored the power of mirrors to concentrate the sun's energy. In the experiment Shuman loved to duplicate, Kircher reflected the sun's light onto a patch of shaded wall from one, two, three, four, and five flat mirrors, each a foot square, from a distance of 100 feet. He tested the heat of the combined reflections by putting his hand in the patch of reflected light on the wall.

The heat from a solitary mirror, he reported, was cooler than the sensation of direct sunlight on the skin. The heat from two was somewhat warmer than direct sunlight. The combined light from three mirrors felt on Kircher's skin like the sensation of heat experienced near an open log fire. From four mirrors, the heat was only just bearable. And from five mirrors, the heat was such that Kircher could leave his hand in the searing spot of light only moments before pain forced him to withdraw it. Kircher recorded in his account:

> *Thence I concluded that by multiplying the mirrors and giving them suitable direction, effects would be produced not only more intense, but concentrated at a greater distance. Five mirrors did so at a distance of 100 feet. What terrible phenomena might not be produced if, for example, a thousand mirrors were so employed! I earnestly pray mathematicians to try this experiment with great care. It will be found that there is no apparatus so well calculated to cast ignition to a great distance.*

By adding more mirrors, Kircher thought he could create a solar beam of concentrated rays so intense that it would cause wood to burst into flames. From these experiments Kircher concluded that if one wanted to burn something at a distance, a large number of plane mirrors, properly arranged so they all shone on the same spot, would be more practical than a single, curved mirror.

This was the method Father Kircher believed Archimedes had employed at Syracuse, in another early use of solar energy—perhaps the most famous one—that left a profound impression on the young Frank Shuman. In an era of difficult travel, Kircher had made the long trip to Sicily so that he could personally inspect the site of Archimedes' reputed solar war machine.

Archimedes' War

The legend of Archimedes and his solar mirror (for many scientists and historians believe it to be no more than that—a legend) is one that modern writers have seen as an eerie foreshadowing of Strategic Defense Initiative (SDI)/Star Wars technology. It, too, fired Frank Shuman's imagination—a tale that touched on his fascination with solar energy as well as his interest in military technology.

It is likely that Shuman first came across this story in a primary school history of Greece and Rome popular after the Civil War. In it was a brief account of the exploits of Archimedes in the Second Punic War (in the third century B.C.) and the suggestion that the sun may have been used as a military weapon to repel an invasion of Sicily by the Roman consul Marcus Claudius Marcellus.

A number of variants of the story exist, detailing just what Archimedes is reputed to have done. In 1200, the Byzantine writer John Tzetzes (not regarded as a particularly trustworthy historian) gives us this colorful account in the second volume of his *Book of Histories*:

> *When Marcellus withdrew them [his ships] a bow-shot, the old man [Archimedes] constructed a kind of hexagonal mirror, and at an interval proportionate to the size of the mirror he set similar small mirrors with four edges, moved by links and by a form of hinge, and made it the center of the sun's beams—its noon-tide beam, whether in summer or in mid-winter. Afterwards, when the beams were reflected in the mirror, a fearful kindling of fire was raised in the ships, and at the distance of a bowshot he turned them into ashes. In this way did the old man prevail over Marcellus with his weapons.*

As Athanasius Kircher would later show, an array of flat mirrors such as Tzetzes describes would have been much easier to build than a giant parabolic mirror, though it would have had the same effect.

Fanciful image of Archimedes using a burning mirror to defend Syracuse Harbor from Roman galleys in the Punic Wars. Whatever the truth of the legend of the Burning Mirrors of Syracuse, the scene shown is almost certainly not what took place.

Zonaras (another twelfth-century chronicler with a shaky reputation) adds:

> *At last in an incredible manner he burned up the whole Roman fleet. For by tilting a kind of mirror toward the sun he concentrated the sun's beams upon it; and owing to the thickness and smoothness of the mirror he ignited the air from the beam and kindled a great flame, the whole of which he directed upon the ships that lay at anchor in the path of the fire, until he consumed them all.*

The historian Lucian gives us an earlier rendition, about a century after the events allegedly took place. The tale crops up also in the memoirs of Galen, better remembered with Hippocrates as one of the great early physicians. Galen speaks of it in passing: "In some such way, I think Archimedes, too, is said to have set on fire the enemy's triremes using flammable material."

Galen's hedging suggests he may have had doubts. Oddly, he specifically mentions "triremes" (vessels with three men to an oar) when it is generally accepted that quinqueremes (five men to an oar) formed the backbone of the Roman navy during the Punic wars. Lucian says only that "the former [Archimedes] burned the ships of the enemy using his science."

The broader context of the story is this: Toward the end of the third century B.C., the Greek colony at Syracuse had allied itself with the nearby city of Carthage (near modern-day Tunis) in its battle against Rome in the second of the three Punic wars. Syracuse was the largest city on the island of Sicily. In the spring of 213 B.C., Syracuse turned decisively against Rome. The Roman authorities ordered Marcellus, fresh from his victory against Hannibal at Nola (near modern-day Naples), to sail 200 miles south to launch a marine assault on the seawalls of Syracuse harbor.

The most famous citizen of Syracuse at that time was Archimedes (287–212 B.C.), the greatest mathematician of antiquity, discoverer of the ratio between the circumference and diameter of the circle and of the concept of specific gravity. Archimedes had studied with Euclid in Alexandria but as an adult had settled in Syracuse. Although we think of Archimedes today as a philosopher and mathematician, his fame among his contemporaries rested mainly on his military inventions. He now enlisted these weapons to defend his city from the naval attack.

Powerful catapults hurled stone missiles and spears out across the harbor at the invaders. When the vessels came near enough to shore, huge cranes dropped boulders on them or snatched their bows from the water with iron grappling hooks, swinging them on the rocks to be smashed. Once they were beached, giant rams suspended from A-frames battered enemy vessels into kindling.

To his horror, Consul Marcellus learned how far Archimedes had developed the art of military defense. Archimedes had determined (by calculations that have survived) the calibration required to launch missiles of varying weights to specific distances. He deployed boulder-throwing and arrow-firing catapults and even a kind of quick-firing weapon. One of his arrow-shooting catapults could hit a lone sailor at 100 yards, or a group of sailors huddled together on the deck of a ship at twice that distance. His stone-throwing catapults were effective at up to 200 yards, with the heaviest projectiles retaining precision even at that distance.[1]

What a Roman galley in Marcellus's time looked like is still being debated. It is probable they were quite massive, each vessel manned by 300 rowers with a marine landing force of 120, reduced to 80 first-line troops if the "corvus" or ram were employed on the vessel. Including officers, there might have been upwards of 500 men on each ship bombed by Archimedes' catapults or upended by his grappling hooks to be sent crashing into the sea. Casualties would have been high.

Marcellus's sailors and soldiers were terrified. The technology was so new they did not know what to expect next. "If they saw only a rope or a piece of wood," Plutarch reported, "extending beyond the walls, they took flight, exclaiming that Archimedes had once again invented a new machine for their destruction and death."

By all accounts, the resistance he met at Syracuse shocked Marcellus. He withdrew his forces and set up a blockade that lasted 30 months. In the interim, Roman spies infiltrated Sicily and bribed native supporters of the Greeks, who betrayed them to the Romans. Marcellus then launched a successful offensive in 211 B.C., took Syracuse, and sacked it. Upon entering the city, he summoned Archimedes, eager to talk to the 73-year-old philosopher whose military innovations had thwarted him. They never met. According to the story recited by generations of schoolchildren, Archimedes was slain by a Roman soldier—allegedly because he was too preoccupied with a problem of geometry diagrammed in the sand to obey the centurion. When the Roman stepped on his lines, Archimedes shouted, "Don't disturb my figures!" Outraged, the soldier impaled the old man on his sword.

---

[1]Based on accounts of these devices, British engineers built a facsimile in England in the nineteenth century that could hurl a 51-pound ball 300 yards.

Though the Greek scientist had done everything possible to defeat him, Marcus Claudius Marcellus built a tomb for Archimedes. Cicero records that he visited Sicily about a century later and found the monument neglected and overgrown with thorn bushes. It has since disappeared.

Testing the Legend

The accepted accounts of the siege of Syracuse say little or nothing about burning mirrors. So where did the story come from? And what truth, if any, is there to the tale?

Analysis of the siege of Syracuse has a venerable history. Every generation of historians and engineers seems challenged to prove or disprove the account of the burning mirrors, and the balance of opinion has swung back and forth over the last 500 years. At the end

A large, lightweight German burning mirror of the late 1700s being used to set fire to a pile of wood at a distance of about 30 feet. This compound parabolic mirror used scores of flat pieces of thin brass plate nailed onto a parabolic armature or frame made of wood. Mirrors of this type, often 10 feet or more in diameter, were by far the most powerful solar reflectors yet developed and could focus the concentrated rays of the sun on a target area less then 1 inch in diameter. Wood burst into flame almost instantly. Copper ore melted in 1 second, lead in the blink of an eye.

of the seventeenth century, belief changed to skepticism as the actual properties of reflectors became better known. In the eighteenth century, belief overcame skepticism after the experiments of Kircher and Buffon, who showed it was possible to light fires at a distance with lenses and mirrors. In the nineteenth century, support for the story was still widespread as the limits of technology were unbridled by new scientific discoveries—though classical historians continued to express doubt based on textual studies. A period of renewed credulity around Frank Shuman's time gave way to doubt in the first part of the twentieth century. More recently, the story enjoyed a scientific revival in the 1990s.

It is settled fact that handheld concave mirrors made of polished brass were used by the Greeks and Romans to "draw down from the skies the pure and unpolluted flame from sunbeams" (in Plutarch's words), to kindle sacred fires in holy places—for example at the Temple of the Vestal Virgins at Delphi or in the temples in Rome where a sacred fire was usually kept alight near the altar. Burning mirrors were especially valued when a sacred flame sputtered out and had to be relit. In a special ceremony, temple acolytes would direct focused sunlight onto "dry and light matter." Set afire by the concentrated solar flux, the burning tinder would be taken inside the temple to rekindle the fire. The sacred light of the sun was thus transported into the temple and preserved there.

A little before the time of Archimedes, the Greek mathematician Diositheius described how a parabolic mirror can bring parallel rays of sunlight to a point focus and reported that such a mirror generally produced higher temperatures than a simple spherical mirror. Diositheius is said to have built the first parabolic mirror (albeit a small one) about a century before the siege of Syracuse—that is, in the fourth century B.C.—and his work is cited by Diocles in his treatise, *On Burning Mirrors*, which contains the first geometric discussion of the focal properties of spherical and parabolic mirrors.

No eyewitness accounts of the siege of Syracuse have survived. The best known chroniclers of the Second Punic War are Polybius, Livy, and Plutarch. All three give detailed and similar accounts of the part played by Archimedes in the city's defense. All three emphasize the devastating effect his mechanical weapons had on the Roman assault, leading Marcellus to abandon the idea of a direct attack in favor of a protracted siege.

Polybius is probably the most important authority. His histories were written within the lifetime of the combatants and he made a practice of interviewing witnesses himself (though it is not clear he did so at Syracuse). He was also especially interested in the technical dimension of war. Polybius makes no mention of burning mirrors.

Plutarch's silence is also damning. He wrote of burning mirrors in other works, and his essay *On the Face of the Moon* shows that he was familiar with geometrical optics. He could not have failed to mention burning mirrors at Syracuse had he heard of them, especially since he greatly admired Archimedes, whom he tended to lionize.

Modern writers who have dismissed the burning mirrors as myth suggest a possible confusion with Greek fire, a mixture of substances variously described as pitch, naphtha, sulfur, and charcoal that could be thrown in pots, launched from catapults, or discharged from tubes. Archimedes certainly had the technology to launch pots containing flammable materials at naval targets.

Fire pots were more likely than a burning mirror to start a shipboard conflagration because the energy contained in gobs of burning pitch is far greater than the energy that could be concentrated with a mirror. And catapults could have launched fire pots at their targets accurately. Their trajectory would have been independent of the position and angle of the ships relative to the sun, the time of day, and the weather, all constraints with burning mirrors.

Burning pitch could even have been made to fall inside the galleys, where the rowers and fighting men were crowded together. Blazes ignited by Greek fire could only be extinguished with vinegar or wine, whereas seawater could be used to extinguish any ordinary fire set by a lens or mirror. If Archimedes had wanted to use fire as a weapon at Syracuse, a good case can be made for his choice of fire pots rather than burning mirrors, had both been available to him.

And yet, the use of mirrors to burn and injure enemies is a theme that recurs in ancient chronicles. It is described by the sixth-century Greek geometer, Anthemius of Tralles (the Lydian aristocrat who is better remembered as the architect hired by the Emperor Justinian to rebuild St. Sophia in Constantinople, giving it its modern form). Anthemius suggested that a number of small flat mirrors arranged to reflect the sun to a single point could be used to blind an enemy and, if conditions were just right, to ignite clothing. He went on to say that he had assembled just such an array of mirrors to test the design and to demonstrate that the focal point of the mirrors could be mechanically adjusted to track a moving target, such as a ship:

> *Combustion will be caused more effectively, if fire is produced by four or five mirrors, or even as many as seven, and if they are distant from each other in proportion to their distance from the point of combustion in such a manner that the rays overlap each other and produce the desired heating more intensely. Moreover, it is possible to blind the sight of an enemy by the construction of these same plane mirrors, because when the enemy advances, he does not see the approach of his adversaries, who have plane mirrors fitted to the upper parts, or to the insides, of their shields, so that the sun's rays are reflected to the enemy . . . and they are thus easily blinded and routed.*

Anthemius states unequivocally that Archimedes used burning mirrors: "Archimedes cannot be deprived of the credit of the unanimous tradition which says he burnt the enemy fleet with the rays of the sun."

A number of later analysts concurred, including Athanasius Kircher. Ever the experimentalist who wanted to see things for himself, Kircher was so interested in the burning mirrors of Archimedes that he undertook a journey from Rome to visit Syracuse harbor around 1645. Kircher's earlier experiments had already demonstrated that a concave mirror could carbonize wood at 15 paces. While the ancient chroniclers had talked of various distances between 3000 paces and 200, Kircher's walks along the coast led him to conclude that the galleys of Marcellus could not have been farther than 30 paces from Archimedes' men when they were set aflame. Though skeptical about so much else, Kircher subscribed completely to the story.

Recent Analysis

Though no modern historian believes Archimedes used a single giant lens or mirror as a weapon, a more plausible version of the tale substitutes a collection of small mirrors, likely in the form of polished shields held by ranks of soldiers, for the single mirror. In this scenario, as the sun approached its zenith in a cloudless sky, the Roman galleys skirted the shore preparing to land their forces. As the enemy ships approached within "a bow-shot," Archimedes gave a signal and a few hundred soldiers standing along the waterfront raised their highly polished bronze shields to reflect the sun against a target on the sails of the nearest ship. The concentrated heat from these shields, acting as mirrors, was so intense that the dark sailcloth smoldered and exploded into flames within seconds. Panicked, the Roman fleet retreated, giving the Greeks time to strengthen ranks and foil the invasion.

Though still implausible, this interpretation gave rise to a debate among physicists in the mid-1980s about the technical requirements of such a defense. In 1992, two British scientists at Leicester University published an analysis of the physics involved. They too cast aside the notion of a single giant lens or mirror as beyond the technical resources of the era. Experimenting with sheets of metal on plywood, they found that the biggest reflector a soldier could comfortably handle was about 3.5 feet high by 2.5 feet wide—roughly a yard square in area—with its weight resting on the ground through a spike or pivot, in the manner of a double bass.

Given the geography of Syracuse harbor, the British scientists determined that the Roman attack must have come from the east. The noon sun around the time of the spring equinox would have been due south at an altitude of 53 degrees. The "length of a bowshot," specified by many of the accounts, could be anywhere between

50 and 100 yards (roughly the distances used in modern archery competitions) but because ancient bows may have been less powerful, the British team took "a bow-shot" to mean 50 yards.

The Leicester University investigators conducted field tests using flat bronze reflectors. Given the intensity of sunlight at Sicily's latitude in the spring, they determined that temperatures of 900 to 1100 degrees Fahrenheit could easily be produced on a nonreflecting target 50 yards from shore if enough (a few hundred) properly oriented mirrors were focused on it. Such a concentration of solar energy would have been more than enough to ignite the maroon-colored sails of Marcellus's fleet in seconds. However, the researchers noted that Roman galleys traditionally furled their sails in battle or when approaching shore, so setting the sails alight might not have been an option.

Alternatively, the Greeks could have attacked the wooden hulls of the Roman ships directly. This would have been harder, but not impossible. The British team calculated that the combined efforts of 440 men, each wielding a metal shield functioning as a mirror about 1 yard square, could just begin to ignite a damp patch of wood half a yard square at a distance of 50 yards. The resulting fire would have been an unspectacular affair—nothing like the explosive ignition described by Zonaras—involving smoldering and charring, with the eventual ignition of the wood into flame. The Romans could easily have extinguished such a fire with buckets of seawater.

Based on this analysis, the British scientists concluded that the legend of Archimedes and the burning mirrors was a myth—if only because Archimedes would not have approved such a poor use of manpower. That did not rule out, in the eyes of these investigators, the use of burning mirrors altogether. Far more vulnerable than the wooden hulls of the galleys were the Roman soldiers assembled on deck, and especially the officers usually massed at the stern. Using numbers developed by a Japanese study of fires and firefighters designed to measure the maximum heat that humans can briefly withstand, the British showed that burning mirrors sufficient to ignite a wood hull produce a heat flux that exceeds this maximum by a factor of 26—creating, from the perspective of the Roman seamen, a blazing inferno of sunlight that would have sent even war-hardened veterans fleeing in a panic of blindness, terror, and pain.

The study concluded that if Archimedes had targeted the Romans themselves rather than their vessels, only 10 percent of the flux required to ignite wood would have been effective, requiring a more manageable force of 50 men with mirrored shields rather than the 440 needed to ignite wood planking. An extrapolation showed that the superimposed beams from 100 men with shields would have been equally devastating at 100 yards. This is the length of a football field, a significant distance. Such a weapon would have dazzled and confused the Romans and might have been used in conjunction with pots of Greek fire.

Although the British team discounted the use of burning mirrors at Syracuse to start fires aboard Roman ships, they thought it at least plausible that a "mirror platoon" of 50 men could have selectively inflicted nasty burns on targeted steersmen, officers, or Marcellus himself (his uniform would have made him stand out)—in this way creating a more personal motive for his retreat.

And yet it is hard to believe that if burning mirrors had been effective—in setting fire to ships or to blind or burn personnel—they would not have become a more popular weapon of war after Archimedes' time. In the centuries of almost continuous warfare in Rome and Byzantium that followed the siege of Syracuse, functioning burning mirrors would have been a fabulous weapon to supplement Greek fire. Once Anthemius had designed a working prototype, as he records he did in his chronicle, why didn't his patron, Justinian, command him to build a full-scale military model?

Painstaking review has convinced most recent analysts that the historical evidence for Archimedes' burning mirrors is feeble and contradictory. Some of the later historical authorities, like Zonaras and John Tzetzes, are themselves unreliable, while the more trusted contemporary authorities are silent—not a combination to inspire confidence. But whatever its truth, the story of the burning mirrors of Syracuse supports the notion that the Greeks understood, at least conceptually, that sunlight contains vast amounts of energy—dangerous amounts, one might say—and that this energy could be concentrated and intensified with mirrors, an important fundamental insight that would soon be developed more fully in Europe.

Experiments in North Africa

From the end of the Greek and Roman period, contemplation of the sun as a source of power declined as Europe moved into the Dark Ages, and we hear little more of lenses, mirrors, and solar power in Europe until the Renaissance. Yet while Europe slumbered, science was flourishing in the Arab world. Solar experiments were popular in the sunny regions of North Africa, where the first contributions to the theory of optics in 800 years were developed by Abu Ali al-Hasan al-Haitham.[2]

Al-Haitham is best remembered today for having refuted the Greek theory of vision, which held that "visual rays" emanate from the eyes onto objects observed like the beam of a flashlight. He correctly argued that vision results from light entering the eye, not the reverse. To prove his point, he dissected mammalian eyes, describing their lenses and operation.

Born in 965 at Basrah, a port city in what today is southern Iraq, al-Haitham studied Greek and Roman texts at Cairo, where he spent much of his life. Though he regarded himself primarily as a mathematician, he tackled practical problems like the origin of rainbows, binocular vision, the apparent increase in size of planets near the Earth's horizon, reflection, refraction, focusing with lenses, and the properties of spherical and parabolic mirrors.

Al-Haitham was excited by Diocles' discussion of burning mirrors, which he called "one of the noblest things conceived by geometricians of ancient times." In particular, he was intrigued by a misconception that was prevalent in much of the Greek literature on the subject (and which would crop up again in twelfth-century Europe): the idea that

[2]His name, later Europeanized by Roger Bacon, is sometimes recorded as "Alhazen."

it was possible, in theory and perhaps even in practice, to create a "perfect" burning mirror that would bring the rays of the sun to an infinitesimally small point of focus, which would then reach an infinitely high temperature. The thought was that such a mirror might make a devastating weapon. It was an appealing concept.

The goal is stated nicely by Diocles in his treatise *On Burning Mirrors*:

> *Pythian the Thasian geometer wrote a letter to Conon in which he asked him how to find a mirror surface such that when it is placed facing the sun the rays reflected from it meet the circumference of a circle. And when Zenodorus the astronomer came down to Arcadia and was introduced to us, he asked us how to find a mirror surface such that when it is placed facing the sun the rays reflected from it meet a point and thus cause burning.*

On Burning Mirrors goes on to show that Diocles understood correctly that the creation of a dimensionless point was impossible because the focus of light rays in a lens or mirror always has area. Although later Greeks ignored Diocles' text, it was a major influence on Arab mathematicians, including al-Haitham. Al-Haitham was also familiar with the optical theories of Anthemius of Tralles, with whom he concurred that Archimedes had understood enough about mirror optics to have used burning mirrors as a weapon.

A Weapon for the Crusades

In a wonderful serendipity, European scientific knowledge preserved in a distant part of the world flowed back into Europe in the late Middle Ages, the continent from which it had utterly disappeared in the Dark Ages. Al-Haitham's Arabic writings—including *On the Light of the Moon*, *On the Halo and the Rainbow*, *On Spherical Burning Mirrors*, and other works on light and optics heavily influenced by Greek thought—were translated into Latin by Gerard of Cremona around 1270. Though al-Haitham's concepts had not created much of a stir in Cairo, they were viewed as revolutionary in

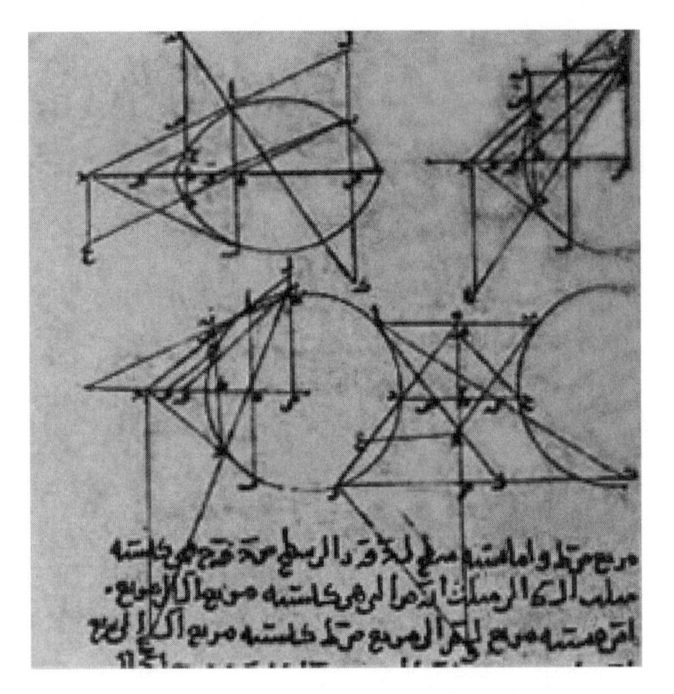

Diagrams from Al-Haitham's studies of optics in an Arab manuscript.

Europe, where Aristotle's worldview had stifled all competing ideas. Copies of the translated texts soon found their way to the University of Paris, where stories of the technical exploits of Archimedes were also being retold.

Several of al-Haitham's books fell into the hands of Roger Bacon, a wealthy Oxonian lecturing in Paris at the time. Burning mirrors exerted a peculiar fascination over Bacon, particularly as weapons of war. He had read Diocles and Anthemius of Tralles and was generally familiar with the concept of concave mirrors. But he was captivated completely—spellbound, really—by the Latin translations of al-Haitham's works, which dealt with burning mirrors with a degree of technical precision previously unknown, providing detailed instructions on their construction.

Throughout his long life (he would live to be nearly 80 at a time when life expectancy was under 40), Bacon was a tireless campaigner for experimental science, and the experimental foundation of al-Haitham's work appealed to him. Bacon had no reservations about accepting the story of Archimedes' use of burning mirrors at Syracuse, and he feared that the technical improvements discovered by Arab scientists posed a grave threat to Europe's Christian kingdoms. In a letter to the Vatican that has survived, Bacon warned that Al-Haitham's deadly innovation in the construction of mirrors was

> *to fashion a mirror that all the rays falling on its whole surface may converge to a single point, and what is more, at any distance from the mirror desired. This is the ultimate that the power of geometry can do. For this mirror will burn fiercely everything on which it is focused. We believe that the Antichrist will use these mirrors to burn up cities and camps and armies. If a moderate convergence of rays by refraction or by a concave mirror burns perceptibly, how much more so without limit, when rays without number converge by means of this mirror?*

Though he should have known better, Bacon had fallen into the fundamental error of the ancient Greeks, an error that would not be corrected in Europe for several decades. Diocles had shown, and al-Haitham had confirmed, that the power of any mirror or lens to concentrate the solar flux is proportional to the area of its light-receiving surface. The notion that attracted Bacon was therefore a chimera: that al-Haitham's mirror could concentrate the sunlight reaching the area of the mirror to a single, dimensionless point of infinite temperature.

Because the sun presents to an observer on earth a disk, not a point, its light can never be focused to a dimensionless point. Instead, the focus of the rays always creates an image of the sun that covers an area proportional to the size of the sun's disk, a fact that effectively limits the concentration of solar heat in proportion to the light-gathering area of the lens or mirror. The larger the area of the lens or mirror, the greater the amount of the sun's heat that can be gathered.

In spite of this theoretical error, Bacon's three principal works, all written for Pope Clement IV, argued passionately in favor of preferring experiment to doctrine, no matter how lofty or ancient the source. To Bacon, al-Haitham's work on the heat-concentrating powers of mirrors seemed to provide practical clues about how Muslims might respond militarily to the Crusades. As he warned Clement, al-Haitham had provided a blueprint for building a doomsday weapon that might any day be wielded by what Bacon called "the Antichrist"—by which he meant the Muslim armies the Crusader knights had encountered in the Holy Land.

Bacon exhorted Rome that "an enemy Saracen shows in a book on burning mirrors how this instrument is made" though he admitted that the author had not unveiled the finer points of its design. He speculated that these were contained in another treatise, one that had not yet been translated from Arabic, possibly because its Muslim owners were averse to seeing it fall into enemy hands.

Bacon was himself interested in performing experiments with burning mirrors, and in that sense his correspondence with Clement may have been little more than an early example of a grant proposal—an effort to persuade the pontiff to support financially a new technology that could serve the interests of the Vatican in war. It is surely also an early example of the use of empirical science, in preference to metaphysical speculation, as a basis for invention, even if only in the pursuit of military ends.

Clement had befriended Bacon before he became pope. While he took a personal interest in Bacon after ascending the throne, even commissioning several of his books, Clement elected not to fund Bacon's experiments. Rebuffed, Bacon assured the pope that he would carry on without official support, and that the "most skillful of Latins is busily engaged in the construction of such a mirror." Bacon later asserted that he and his colleagues had finished building the mirror after working "many years . . . at great expense and labor . . . abandoning studies and other necessary business." To demonstrate the utility of these efforts, Bacon claimed that if Christians "living in the Holy Land had twelve such mirrors," they would have the means to "expel the Saracens from their territory, avoiding any casualties on their side." This, he told Clement, might obviate the need for Crusader reinforcements, a drain on the purses of Europe's monarchies and a looming financial disaster that was already a papal concern even at this early period of the Crusades.

Apart from his letters, no evidence that Bacon ever built a solar weapon or had access to one has ever turned up. For such a device to have been effective at the range of the spears, arrows, slings, and catapults of the thirteenth century, it would have had to have been truly colossal. That alone argues against it, for the same reasons Archimedes would likely not have had use of such a device.

Meanwhile, Bacon ran afoul of the conservative cardinals in Clement's court who envied Bacon's personal relationship with the pope and were determined to discredit him. These clerics attacked Bacon, complaining that in his eagerness to drum up papal support, he had failed to reveal the potential limitations of mirrors as weapons. Muslim generals, the cardinals argued, would be thoroughly conversant with optics. Were solar mirrors to be deployed against them, they would certainly attack facing the sun, making it impossible for the mirrors to reflect solar rays in their direction. The Muslim armies, they told Clement, might also choose to move against the Christians on an overcast day or at night, rendering Bacon's solar weapons impotent.

Besides the pope, Bacon had few allies in the papal court. In failing health, Pope Clement IV died on November 29, 1268, in exile at Viterbo, north of Rome. He had reigned less than 4 years. The papacy remained vacant for the next 3 years while the clique of conservative cardinals who loathed Bacon held sway. Unable to agree on a successor to Clement, they were unanimous in their hatred of Bacon. To these churchmen, the empiricism of Bacon threatened the worldview of the Catholic Church, a perspective synthesized by Thomas Aquinas that melded the metaphysics of Aristotle with divine revelation. In this world order it was revered texts, not experimental science, that held final authority. Ideas contrary to these texts were not only an affront to truth, but to God. The idea of transforming the beneficent rays of the sun into a fierce military weapon to incinerate human beings was seen as a deviant and evil idea, the work of witchcraft and the devil.

With Clement no longer available to protect him, Bacon was exposed to the unbuffered wrath of his powerful enemies. Sometime between 1277 and 1279, he was condemned to prison by his fellow Franciscans because of "suspected novelties" in his teachings. His fall from grace surprised no one. Bacon had been voluble in his attacks on the theologians and scholars of his day and seems to have gone out of his way to provoke them, not just though his insistence on experimentation. He was imprisoned 14 or 15 years, no one is sure exactly how long. Though he pursued his scientific studies after his release, he was careful to do so in secret.

Leonardo da Vinci's 4-Mile Mirror

So powerful was the lesson of Bacon's condemnation that for 2 centuries after his suppression European thinkers were silent on the subject of burning mirrors. Early in the sixteenth century, Leonardo da Vinci, a man who, like Bacon, was also not circumspect when it came to the Church, proposed a fabulous concave mirror 4 miles in diameter to be built in an excavated bowl-shaped recess in the ground. Leonardo's notes suggest that the huge stationary mirror would be dedicated to peaceful uses—a source of heat and power to run commercial enterprises, not as a weapon of war.

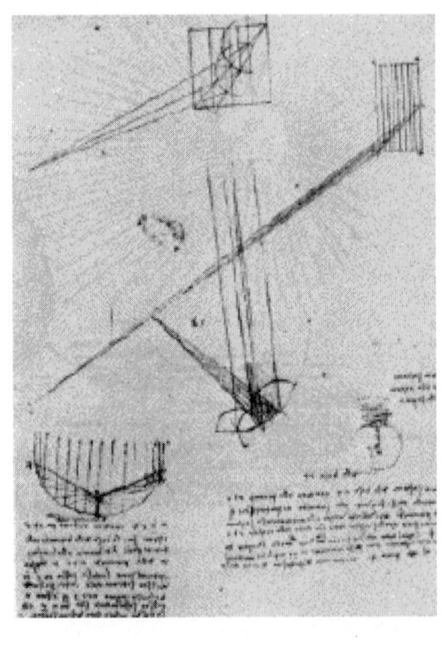

Leonardo Da Vinci and his drawing of a 4-mile burning mirror to be set in the ground.

The story of the giant mirror came to light in the 1970s in a textual analysis of Leonardo's surviving papers by a scholar at the University of California, Carlo Pedretti. Pedretti determined that Leonardo's experiments with parabolic mirrors probably began in the first decade of 1500, when he was already in his late forties and early fifties. For Leonardo, this was a time of exhausting travel in search of new patrons who could guarantee him the freedom to study while restoring the material comforts he had enjoyed while working for Lodovico Sforza, his previous patron, who had treated him generously.

Leonardo's need for money led him to accept, in the summer of 1502, an invitation from Cesare Borgia, son of the dissolute Pope Alexander VI, to join him as architect and general engineer in the military campaign intended to give Cesare, with papal-paternal support, a vast personal dominion in central Italy. Whatever reservations he may have had about him, Leonardo traveled with Cesare through the Umbria and Romagna, inspecting fortresses and proposing systems to defend them better. He took part in the capture of Urbino, where he was enchanted by the literary treasures collected by the Montefeltro family in their famous library, which Cesare promptly confiscated and sent to Rome. Leonardo later recorded his elation at finding copies of scientific texts he had never seen before in his life:

> *There is a complete Archimedes in the possession of the brother of Monsignor of Santa Giusta in Rome. The latter said that he had given it to his brother, who lives in Sardinia. It was formerly in the library of the Duke of Urbino, and was carried off from there in the time of the Duke Valentino.*

Among these Archimedean manuscripts, Leonardo found an exhaustive account of the burning mirrors of Syracuse, a story he may have known previously only in sketchy outline. This document provided an impetus for him to perform experiments of his own with models of giant concave mirrors made up of a mosaic of flat pieces of mirrored glass glued to the bottom of a bowl. He was especially interested in the "varnish" used to silver concave mirrors to improve their reflectivity, and his notes contain calculations to quantify the degree to which the sun's rays could be concentrated.

Regrettably, details of the huge in-ground mirror are scarce. The main evidence for it consists of notes from 1513 through 1516 on Leonardo's favored thick blue paper, along with a diagram of the mirror itself. The drawing leaves no doubt that the mirror was to be of enormous size. Just how it was to work is less clear. Leonardo's plan apparently was to create a shallow, concave basin on a flat plain

A ship being set afire by a burning lens set atop a lighthouse in medieval times. Such devices were more imagined than real.

near Rome. At the time Leonardo was on loan to Giuliano de Medici, who had asked him, as a skilled hydraulic engineer, to advise him on one of Giuliano's pet projects—the draining of the Pontine marshes. Leonardo occupied his spare time in the mosquito-filled swamps with the study of optics, diagramming a number of concave mirrors. That work apparently led him to the larger project.

We know that the enormous, man-made depression to hold the mirrored exterior was to be 4 miles wide. And we know that the smooth surface of the basin was to be paved with reflective materials that would concentrate and focus the sun's rays on a central axis, where a tall pole would hold the materials to be heated. The focal point of the giant mirror was to be "four braccia" (about 13 feet) above the deepest, central point of the basin, at the top of the pole. In an accompanying note, Da Vinci wrote:

> *With [this device] one can supply heat to any boiler in a dyeing factory. And with this a pool can be warmed up, because there will always be boiling water.*

Leonardo also saw potential for this device to melt metals, and he included a kind of drain or opening at the bottom of the basin through which molten materials could flow. He wrote:

> *In order to melt the object [at the mirror's focal point] we must have its support above the bottom of the "vaso igneo" [that is, fire dish] where it will be reached by the power of all the rays converging from every direction.*

Like so many of Leonardo's astonishing conceptions (including his proposed flying machines and the submarine he designed to travel beneath rivers), this one was likely not technically feasible in its time. Carlo Pedretti has suggested that a whole subset of Leonardo's research was focused not so much on actual projects for realizable devices, but as a sort of technological dream, the product of his own special imagination. The texts accompanying these designs often had an oracular tone and were accompanied by reflections on human nature, as in some of his studies of flight or in the famous note in which he stated that he wished to keep secret his invention of a way to breathe underwater to prevent its use as an instrument of death. It is surely no accident, in an era that regarded solar technology as having mainly deadly military application, that Leonardo's giant mirror was designed only for peaceful ends.

Though no evidence exists of anyone ever trying to construct Leonardo's vast device, handheld burning mirrors were often used for soldering in the fifteenth century. Leonardo records, in a note dated 1515, that it was a burning mirror that Verrocchio used in soldering together the triangular sections of the copper balls for the lantern of Florence's Santa Maria del Fiore Cathedral.

Leonardo's work on mirrors was a turning point in the history of solar power, and not only for its enormously consequential shift in focus from destructive to industrial uses. Before his time, investigations of solar energy were sporadic, the results anecdotal—and many of the anecdotes likely apocryphal. In the centuries after Leonardo, interest in mirrors, lenses, and solar devices flourished, fueled by con-

An Italian fresco depicts the defender of a seawall setting fire to the hull of a wooden warship with a burning glass. Though such devices were widely discussed in Europe from the time of Roger Bacon to Leonardo, it is not likely anything like this design would have actually worked in practice.

The great solar lens of the Paris Academy of Sciences (1782). This horse-drawn device was constructed at the royal factory of St. Gobain under the direction of the great French chemist Antoine Lavoisier in order to produce intense localized heat to melt ores and perform chemical experiments requiring temperatures higher than could be produced in a refractory furnace. The experimenter wore dark glasses to protect his eyes from the intense light. Benjamin Franklin, who was in Paris as an American diplomat at the time this apparatus was in use, saw Lavoisier demonstrate the extraction of platinum from its ore with the heat of this lens.

tinued experiments that even today we would regard as scientific in their formulation and execution.

These experiments provided the platform of scientific knowledge that supported the explosion of interest in solar power that would occur in the nineteenth and twentieth centuries. They provided also the scientific scaffolding for the revolutionary machines that would arise from the prolific imagination of Frank Shuman.

3

The 4-Acre Solar Machine

When he disembarked the SS *Mauretania* at Liverpool's Albert Dock, Frank Shuman wasted no time sending a telegram to his consulting engineer, A. S. E. Ackermann, whose supportive analyses of the solar power plant in Philadelphia had proved so helpful in lending credibility to his proposals in the American scientific and business communities. He advised Ackermann of his imminent arrival in London and caught the overnight express to London from Lime Street Rail Station. He was at Euston Station in time for breakfast.

Ackermann's shingle as a consulting engineer was hung at 25 Victoria Street, Westminster, an elegant mock-Palladian building overlooking the cathedral. His office boasted one of the earliest telephones installed in the city (Number 244 Victoria). Here, in an airy study paneled in dark wood and comfortably furnished in the heavy rosewood and mahogany pieces, richly upholstered, favored by prosperous London professional men, Ackermann and Shuman began to frame their campaign.

The smooth luster of the brass fittings on Ackermann's desk gleamed through a haze of cigar smoke in the gray light of an early winter as they discussed the finer points of their strategy. They sequestered themselves for several days. Though they had yet to arrive at the exact figure, they knew they needed to raise subscriptions of several hundred thousand pounds to capitalize the Sun

New Jersey

111

THE SUN POWER COMPANY

This Certifies that Frank Shuman — is the registered holder of Four —— Shares of the Capital Stock of THE SUN POWER COMPANY

Transferable only on the Books of the Company by the holder hereof in person or by duly authorized Attorney upon surrender of this Certificate properly endorsed.

SHARES $100 EACH

Witness the Seal of the Company and the signatures of its President and Treasurer this 20th day of Jany 1909

Constantine Shuman, TREASURER

Frank Shuman PRESIDENT

Shares 4

Stock certificate from the Sun Power Company.

Power Company (Eastern Hemisphere), Ltd., the firm they had incorporated in 1910.

If Ackermann's endorsement had proved decisive in generating support for Frank Shuman in the United States, his collaboration in Great Britain was to become even more valuable. Ackermann was exceptionally well connected. In 1898 he had been elected honorary secretary of the Civil and Mechanical Engineers Society, a post he held until 1910 when that group was absorbed by the larger Society of Engineers, on whose board he also served. He then became secretary of the merged organization, a post he would keep until 1938 when he retired. Through these professional posts and his tony consulting practice in Victoria Street, he had access to every scientist of note in the kingdom.

Alfred Seabold Eli Ackermann was a Londoner, like his father and grandfather. Born in 1867, he had been taken as an infant to South Africa, where he grew up in Cape Town. He showed unusual ingenuity as a child, at 17 building a violin he had designed himself, which was exhibited at an industrial exposition in the city. He got his early university training at the South African College, a forerunner of what was to become the University of Cape Town. He was a precocious student. Mechanical engineering had been a part of his life almost from birth. As a teenager, he was articled to his father as an apprentice. Ackermann senior was himself a respected engineer and a stern taskmaster. His reputation helped the young man secure early assignments with Delta Metals; the British Gas Engine Company; and the James Simpson Company, a firm of hydraulic engineers in London, who hired the younger Ackermann when he returned to his homeland in his early twenties. He later asserted that he had seen a greater variety of engineering problems before he was 30 than most engineers confront in a lifetime.

Back in London, he entered the City and Guilds College in 1890 to continue his professional training. He took courses in civil and mechanical engineering taught by Professor William C. Unwin, a man well known on both sides of the Atlantic. Unwin was a member of the International Niagara Commission, a group organized by the American inventor George Westinghouse to consider projects to harness the energy of the thundering falls where the Niagara River drops from Lake Erie to Lake Ontario. Other members of that group included William Thomson (Lord Kelvin), probably the most famous physicist in the world at the time, and Coleman Sellers, a professor of mechanics at the Franklin Institute in Philadelphia who was also a benefactor of Frank Shuman.

While working with Unwin, whose assistant he became, Ackermann spent part of his week at the Priestman Brothers engineering plant at Hull, a center of high-technology engine design, helping to organize trials of the William Dent Priestman heavy-oil engine, a British forerunner of the machine that would make Rudolf Diesel famous in Germany when he patented it there in 1892. Priestman was awarded a John Scott Medal in Philadelphia in 1894, the same year Frank Shuman was awarded his for his wire-glass machine.

Ackermann was among the earliest engineers to develop the concept of performance evaluation. His consulting work focused on the testing and reporting upon all kinds of machinery. He advised clients on the fuel economy of their engines. He was interested in the construction and testing of boilers, whose frequent explosions in the early twentieth century were still an occupational hazard. His work with boilers led him to study vibration, noise, and pressure testing.

Ackermann was still in his twenties when he set up practice as an independent consultant, a decision he made so that he would never be tied to a single commercial enterprise. In 1902 he turned to manufacturing processes and quality testing, becoming an adviser to the Linolite Company, a firm that made electric light fixtures. His work with electricity led him to a study of dynamos and steam engines. He traveled extensively in the United States, where he met Frank Shuman, whose interest in the direct use of solar energy to produce steam he came to share.

At the time of Frank Shuman's visit to London in the fall of 1911, Ackermann was already intimately familiar with Shuman's work. The previous August he had spent several weeks in Philadelphia at Shuman's compound conducting field tests on designs of solar absorbers and low-pressure steam engines. He believed Shuman was at a critical juncture in the development of his inventions. He thought Shuman's future now depended on his ability to take his ideas across an invisible frontier that separated experimental curiosities of great promise from lucrative commercial enterprises. It was a bridge few inventors crossed successfully. To do that, Shuman would need help, and Ackermann's business acumen coincided perfectly with Shuman's realization of his need for it.

Ackermann wanted to open up England's economic and social wealth to Shuman's venture, not just the "factors of production" of traditional economics—capital, labor, land—but the skill, passed on by education and training, of Britain's engineers and financiers. It was a question of building momentum. Steeped as he was in so many projects of the late industrial revolution, no one knew better than Ackermann how vulnerable innovative ideas could be, how they could fail to thrive if just the right resources were not available to them at just the right time.

For example, one of Europe's most inventive minds, Leonardo da Vinci, whose notebooks revealed plans for flying machines and submarines as well as a giant solar mirror, failed to realize almost any of his inventions in sixteenth-century Italy. Why? Because capital, materials, and the necessary skills to develop them had not been available, Ackermann argued. Most of da Vinci's ideas never moved beyond the stage of quill scratches on paper, though they may have inspired the inventions of others.

Two centuries later and closer to home, James Watt might have had a similar experience with his brilliant conception of a new kind of steam engine but for the good fortune of his partnership with Matthew Boulton, who supplied capital, machinery, and the skilled craftspeople required to convert Watt's ideas into salable machines. No invention ever became generally accepted, Ackermann believed, until the economic and social resources were available to sustain it.

Both Ackermann and Shuman must have suspected how high would be the hurdles they faced. Even as early as 1911, the prospects for solar energy had become clouded by a set of firmly held but poorly founded beliefs. Critics articulated many concerns: Solar energy was too diffuse to achieve the power needed by a modern, energy-driven society. It was impractical because it was not available at night or on cloudy days. The equipment needed to run a solar plant was far more costly than a conventional coal-fired boiler. Many of those whose financial backing Shuman would seek had already closed their minds to the idea of solar-powered steam engines. Only an intensive educational initiative had any chance of turning that prejudice around. And yet, both men were confident that once the facts were known, they would convince British investors that solar power was a sound proposition, if not in England, which was too far north to get much sun, at least in her colonies. They were convinced that solar power could replace nearly all consumption of coal in Britain's tropical possessions.

As Ackermann later wrote in a retrospective report on solar power technology he prepared for the Smithsonian Institution in 1915:

> *A great deal of ingenuity was expended in working out the details [of the Philadelphia plant] and it achieved practical success. It also demonstrated that radically different methods must be adopted to place sun power on a commercial basis.*

Solar Energy into Mechanical Power

The model sun-power plants in Pennsylvania provided them with practical experience of immense value. A lesson in the history of science that could not have been lost on any engineer in the first decade of the twentieth century was that the most powerful discoveries of science in the previous 50 years had often arisen, not through the study of phenomena as they occurred in nature, but rather through the study of human-made devices, the products of technology. After all, the operations of machines were simple and well ordered compared to the operations of nature.

For example, it was the steam engine—in which heat, pressure, vaporization, and condensation occur in a simple, orderly, and easily observed fashion—that cried out for the science of thermodynamics. The steam engine gave birth to thermodynamics, not the other way around. In fact, one could say that before the steam engine, the world had *no need* for a science of thermodynamics.

Similarly, in Ackermann's own time, the study of aerodynamics and hydrodynamics arose chiefly because airplanes and ships were being perfected by engineers, not because of a sudden interest in the antics of birds and fishes. And knowledge about electricity came mainly from the study of human inventions, not from the study of lightning.

So too, Ackermann believed, would it be with solar power. If technology was concerned mainly with the study of doing and making things, an understanding of power was necessarily at the heart of technology because it was *power* that provided the ability to make or do anything. Existing solar machines would inevitably result in better solar machines as the science of solar energy kept pace with improvements in engineering and design that were already taking place.

Such weighty abstractions fired the optimism of the two men as they planned their solar venture in Ackermann's Westminster suite.

Beyond the generalities of scientific principles, Ackermann had also developed a number of precise conceptions about how Shuman ought to market solar power in England. The design of a successful solar power plant, one that would attract capital sufficient to launch a new company, would have several hallmarks, in Ackermann's view.

First, it would be sold as a complete package to the consumer, who would buy a working steam engine to provide power for whatever machinery the customer needed to operate—forge, pump, mill—it didn't matter. The product to be sold, in Ackermann's thinking, was inexpensive and reliable *mechanical power*, not solar energy collection. Ideally, plant components would be prefabricated in Europe or America and shipped out in crates, ready to be assembled on the client's site in the tropics. Ackermann recounted the strategy in a long article in *Engineering News* in 1916: "Simple and uniform construction of our sun heat absorber would enable us to build the entire framework at our works, ship it to its destination, and there erect it almost without the stroke of a hammer."

The manufacturer would provide the engineering support to assemble the parts and train local labor to operate the plant. That the steam engine was powered by sunlight, in Ackermann's view, should be of little concern to the customer. He was simply buying a steam engine, and one that would operate more cheaply and more reliably than the coal-fired machine employed by his competitor, enabling him to operate his business at lower cost. Such was the strategy that would create confidence in solar power.

Ackermann's second concern had to do with energy storage. Even on a cloudless day, the amount of heat a sun absorber could take up varied enormously at different hours because of the sun's position. Ackermann's tests in Philadelphia had shown that at 9:00 A.M., heat absorption was only 20 percent of that at noon simply because of the lower angle of the sun's rays, and that by 4:00 P.M. the rate had fallen back to 20 percent of the noon figure.

Startup time was also a problem. The Philadelphia plant, which cooled off overnight, could not develop a sufficient head of steam to begin mechanical operation until about 11:00 A.M. By early afternoon the sun was low enough to cause the machine "to knock off work early in the day," Ackermann wrote, "as though it were a member of a union."

The problem was compounded when one took bad weather into account. Unlike Shuman, Ackermann believed that "until some revolutionary advance is made in the means of storing energy, no sun motor can be a commercial success in any humid climate where the sky is frequently covered with clouds. It is not uncommon in most parts of the globe to have not only whole days without sunshine but also a succession of such days, or sometimes even weeks. Of course a power plant subject to such interruptions of operation would be wholly out of the question," Ackermann wrote. "No gains in economy could offset such irregularity of output." If sun power was to be put to practical use, Ackermann was convinced that some method for storing energy was essential.

His third concern had to do with simplifying construction of the heat absorbers to lower their cost. His technical assessment of the Philadelphia plant, portions of which were published in the April 4, 1912, edition of *Nature*, had taken note of this problem:

> *The pipes in the hot-bed correspond exactly to the pipes in a water-tube boiler of an ordinary steam power plant, but whereas these latter are exposed to a fierce heat of 1,000 degrees to 2,000 degrees F., the pipes in the hot-bed are at a temperature of only 200 degrees to 212 degrees as a maximum. Hence, instead of developing a horsepower with only one or two square feet of heating surface, as in a locomotive boiler, or eight or ten square feet, as in a stationary boiler, a sun engine, built on this plan, would require probably one or two hundred square feet [per horsepower].*

Using Shuman's earlier models, Ackermann had come up with a rule of thumb of 160 square feet of collection area per horsepower of output from the steam engine. As early as 1909, based on the work he had seen in Philadelphia, Ackermann had concluded:

> *It is clear than any construction for absorbing sun heat must be of very low cost and very durable. We must collect the heat from a very considerable area in order to develop much power. For a power plant of 1,000 horsepower, a heat absorbing area of 160,000 square feet or nearly four acres, would have to be provided. Manifestly, any construction spread over so large a space must be simplicity itself or the cost of installation and maintenance will make sun power too expensive for consideration.*

Creating the Prototype

Though it was much larger than any existing steam engine they had tested, Ackermann had believed since 1909 that a 1000-horsepower machine was about the right size for a demonstration project. It was large enough to be taken as a serious machine and yet not so large that the design of the plant could not be directly extrapolated from previous work. Ackermann pictured it this way:

> *Imagine such a plant located on the arid plains of Arizona or of Egypt. A space 400 feet square is graded to an absolute level and rolled hard and firm and is then rendered waterproof by covering it with asphaltum and at the same time given maximum power for absorbing solar radiation, its surface being black. Low walls on the four sides convert it into a shallow tank and long partitions divide the tank into 20 separate compartments, each 20 feet wide and 400 feet long. Between each of these divisions is a narrow space which serves as a walk and also as a gutter to carry off water in case of rain, and also in case of the washing off of the glass [covers] with a hose.*

Shuman had thought of the asphalt bottom back in Philadelphia, and Ackermann considered it a stroke of genius. Asphalt was remarkably cheap and did not have to be laid on very thickly because the surface did not have to bear any traffic, merely the weight of a few inches of water. It was absolutely watertight, would form a tight seal with the pine board divisions channeling the water, did not have to be painted black because it was already so, and required no maintenance. If the ground beneath it had not been graded perfectly level, the asphalt would become semiliquid anyway when heated and form a perfectly flat surface on its own. It seemed like the perfect heat-gathering material for the project.

The 400-foot-square heat collection area was to be covered with the cheapest plate glass available. Because a vast surface needed to be covered, the plan was to purchase precut 16-by-24-inch panes of window glass from a Shuman-owned manufacturer. (Even with

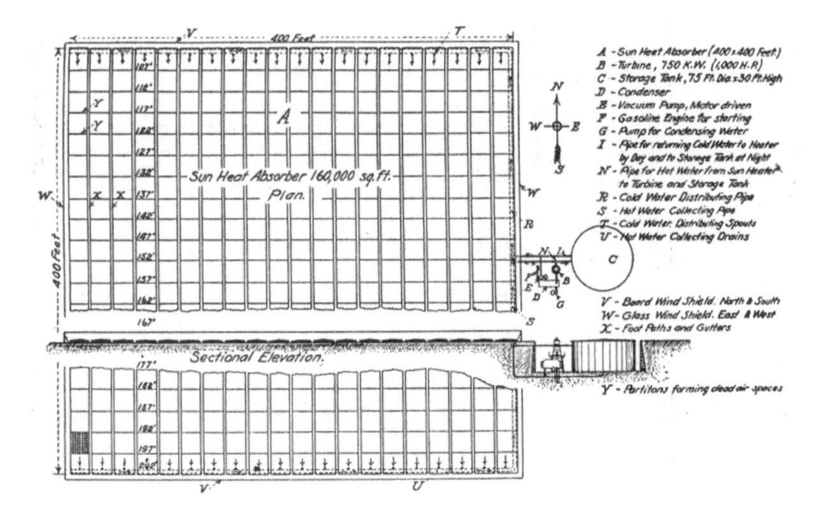

Frank Shuman's 1000-horsepower solar steam engine was to be powered by this 160,000-square-foot "sun heat absorber," an area approaching 4 acres (top). The sectional drawing (center) shows how the hot water would be collected in a 75-foot storage tank set below ground (at right), allowing operation at night and when the sun was obscured by clouds.

transport across the Atlantic, U.S.-manufactured plate glass at the time was less costly and of better quality than that made in Europe.)

Instead of using double panes with a dead-air space between them, as in the experimental plant in Philadelphia, the lower glass pane in the new design was dispensed with and replaced with a thin film of paraffin wax on top of the water. Kept in a molten state by the hot water when the plant was in operation, the paraffin would become completely transparent and would function much as the second layer of glass had in earlier models. Tests in England had demonstrated that paraffin was almost as efficient as a second layer of glass. The small loss in efficiency was offset by three important benefits. Paraffin was far cheaper than glass; it prevented water loss through evaporation; and it cut down by 50 percent the glass surface that needed to be cleaned, eliminating completely the hard-to-reach inner layer.

Ackermann was also counting on the larger surface area of the heat absorber to produce a greater overall efficiency per square foot in heat collection than Shuman had previously achieved. This would occur, he predicted, because a small heat absorber radiates heat from the sides and bottom more quickly than a large one, and because air currents passing over a small area quickly carry the cool atmospheric air to all parts of the glass surface, creating loss by conduction and convection. On a sun heat absorber as large as the one proposed (4 acres), the greater part of the surface would rarely come into contact with cool air. A thermal cushion would protect it, even in a light wind. As Ackermann described it: "The hot air lying above the glass will have a tendency to cling, and the larger the [glass] surface, the more difficult it is to sweep it off."

To minimize further heat loss from moving air, the sun heat absorber was to be protected at the north and south sides by a 10-foot-high corrugated metal fence, and on the east and west sides by a fence of equal height set with glass panes, like a greenhouse. This would allow solar rays to reach the absorber even at sunrise and sun-

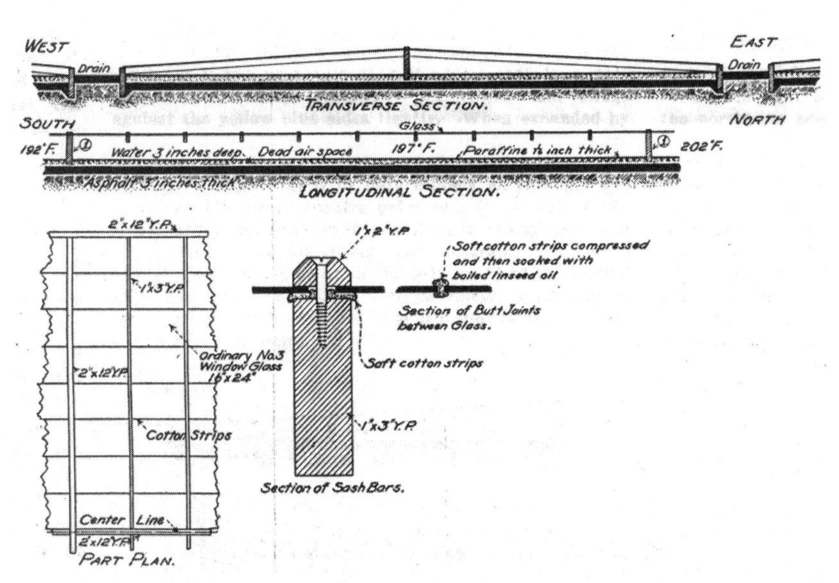

Two insulating layers—one of glass, greenhouse-style, and one of paraffin, floating directly on the surface of the heated water—would help the solar absorber to retain the sun's heat, causing the temperature of the water on a hot day to approach its boiling point. A 3-inch-thick layer of black asphalt, looking much like a giant parking lot, would very effectively have converted the entire solar spectrum to heat, transferring it to the water.

set. The fences, by preventing the rapid movement of air over the glass hot boxes, would tend further to insulate the whole field of glass panes. Under the direct rays of the sun in a hot climate near the equator, a solar absorber so enclosed would quickly heat water to just under the boiling point at atmospheric pressure.

The Cost of Solar Steam

Ackermann and Shuman took great pains to rethink the design of the solar collection area because they were convinced that it would prove to be vital to the economic viability of the whole project. Other elements of the solar steam plant—the engine, the condenser, site construction, and so on—were expected to be at rough parity in cost to a conventional coal-fired plant. It was the absorber that represented the big departure, both in design and in extra expense. Prospective customers would have to be convinced that the extra cost of the absorber represented, in effect, an attractive upfront offset for fuel, because the solar plant would operate without the need to purchase coal.

Though he had not yet attempted a cost-benefit comparison with a 1000-horsepower coal-fed steam engine, Ackermann did calculate that they could build the asphalt-based, glass-covered collection area for $0.25 a square foot, even at remote locations in the tropics. At this rate, the 160,000 square feet of collection area needed for a 1000-horsepower engine would require an investment of $40,000, or $40 per horsepower. Ackermann recognized that such a sum would appear large for what was, in effect, a glorified boiler. He and Shuman would need persuasive evidence that this cost was offset by fuel savings over the life of the plant. It would be necessary, for example, to show prospective buyers that even at unrealistically low imputed prices for coal, the solar plant could still deliver cheaper power and recoup its investment quickly. This was an argument both men thought they could win. Ackermann wrote: "When we recall that this boiler does not require any fuel but collects its own and does its own stoking besides, the price seems low rather than high."

Further analysis would be required to back this claim with hard numbers, but that work could wait. Both men wanted to demonstrate through a simple calculation to potential customers that, for any reasonable price of coal in the tropics, their plant was cheaper and more efficient. Only then would a customer, as a rational investor, be persuaded to commit the larger initial capital a solar plant required. If this argument failed, the rational buyer would opt for the tried-and-true coal-fired alternative for less money and take his chances with coal prices. Most businesspeople preferred to defer payments, Shuman knew, and a stream of coal payments, even if high, stretching years in the future might seem attractively far off compared to the additional out-of-pocket cost of a solar plant. This was true psychologically even if, in the long run, the solar plant proved to be the better bargain. For these reasons, Shuman and Ackermann were both convinced that the extra investment required by the absorber had to be kept as low as possible if prospective investors were not to be frightened away.

Storing Heat

The next problem, heat storage, was not difficult to solve, though it too involved extra cost. The heated water from the absorber was to be piped into a huge tank buried in the ground adjacent to the steam engine. This tank, 75 feet in diameter and 30 feet deep, would hold just under 1 million gallons. For 6 hours of every 24-hour cycle, while the sun was shining at its brightest, the heat absorber would deliver water at 212 degrees Fahrenheit to both the steam turbine and the tank. For the remaining 18 hours of the day, the engine would draw off hot water from the surplus stored in the tank, permitting engine operation round the clock.

Both engineers knew that one of the many remarkable properties of water is that it can store vast quantities of heat. Pound for pound, other materials cannot hold heat nearly as well. Alcohol, for example, has only 58 percent of the heat capacity of water; glass, 20 percent; and steel, merely 11 percent. No substance as cheap and readily available as water can store heat so efficiently. It was the ideal heat storage medium for a solar plant.

A—Sun Heater. (400 x 400 ft.)
B—Turbine, 750 KW. (1,000 HP.)
C—Storage Tank, 75 ft. dia., 30 ft. high.
D—Condenser.
E—Vacuum Pump, motor driven.
F—Gasoline Engine for starting.
G—Pump for condensing water.
H—Well for cold water.
I—Pipe for returning cold water to sun heater during day and to storage tank at night.
K—Pipe for returning cold water from storage tank and turbine to sun heater.
L—Pipe for returning cold water from turbine to bottom of storage tank at night.
M—Pipe for returning cold water from turbine to sun heater by day.
N—Pipe for hot water from sun heater to turbine and storage tank.
O—Overflow from condenser to be distributed for irrigation.
P,P'—Perforated Partitions for equalizing flow of water into and out of storage tank.
Q—Drain from tank pit to well.
R—Cold water sluice to sun heater.
S—Hot water sluice from sun heater.

The low-pressure 1000-horsepower steam turbine (center, on the scaffolding tower) designed by Shuman would convert hot water at sea level into steam by introducing the water into an expansion chamber from which air had been removed. At reduced pressure, the hot water would expand explosively into steam, powering the turbine. A condenser (at left) would redirect the cooled water back into the sun heat absorber. The storage tank (at right) was to hold more than 1 million gallons of heated water, enabling 24-hour-a-day operation of the turbine.

The amount of energy that can be stored in a water tank the size Shuman and Ackermann proposed is very large indeed. A standard measure of energy in use in 1911, as it is today, was the British thermal unit, or Btu, defined as the amount of energy required to raise 1 pound of water 1 degree Fahrenheit. The water in the solar tank when full (992,000 gallons) would have weighed 8.3 million pounds. Assuming the water could be heated to 212 degrees Fahrenheit in the solar hotbed and that it emerged from the steam engine at 102 degrees Fahrenheit, having fallen in temperature by that amount through loss of energy in powering the turbine, the tank could hold 910 million Btu of usable heat energy. Given that it takes 42 Btu per minute to sustain 1 horsepower, this amount of heat could in theory supply a 1000-horsepower engine continuously for well over 2 weeks, assuming all the heat was available.

Of course, all of the heat was *not* available. Even with good insulation, in practice *colossal* quantities of heat would be dissipated to the surroundings. Conduction, convection, leaks, friction in the engine, and radiation from hot surfaces throughout the system would rob the system of heat every moment of the day and night, especially in the cool of the desert. And yet, factoring in liberal figures for heat loss, a tank of the size proposed would still comfortably permit overnight operation of the plant at full throttle—and, with better insulation, even operation during a string of 1 or 2 cloudy days—without interruption. Though a smaller tank would have sufficed for 24-hour-a-day running of the turbine, a million gallons of hot water provided a margin of surplus storage. With such a mechanism, it would be a rare occasion indeed when the steam engine had to shut down for lack of "fuel."

The Solar Absorber

The heat-gathering ability of a 400-foot-square solar absorber was also substantial. The solar constant, the term for the amount of radiant heat reaching earth from the sun, had by 1911 been measured many times by scientists on both sides of the Atlantic.[1] Results varied, because of poor equipment. The most recent measurements had shown that the radiation received per square foot above the earth's atmosphere was about 7.37 Btu per minute. Not all that energy was

[1]The solar constant is defined as the rate at which energy is received from the sun just outside the earth's atmosphere on a surface perpendicular to the sun's rays. It is known today to equal 1.38 kilowatts per square meter (kW/m^2), or 442.4 Btus per hour per square foot (Btu/hr/ft^2), or 1395 watts per square meter (W/m^2).

available, of course. Before the solar radiation reached the surface where it could be put to use, it had to pass through the atmosphere, which absorbed and reflected greater or lesser amounts of it, depending on the level of airborne dust and humidity. If the air was clear and dry, considerably less energy was lost than if the air was loaded with microscopic dust and moisture. Under near-ideal conditions in tropical regions away from coastlines, experiments with balloons and on mountaintops suggested that 78 percent of the solar energy arriving from space made it to sea level. Under these conditions, the Shuman absorber would take in about 920,000 Btu per minute, or the heat equivalent of burning 2.2 tons of coal per hour.

A 1000-horsepower steam engine would require about 61 million Btu of heat to operate continuously for one 24-hour cycle, or the heat equivalent of about 2½ tons of coal.[2] During the 6 brightest daylight hours on a clear, sunny day, from 9:00 A.M. to 3:00 P.M., the absorber would take in about 331 million Btu of heat from the sun, or five and a half times the energy needed to run the turbine for a day. The 270 million Btu daily surplus (on days in which the sun was shining) would be stored as heat in the water tank for later use, or lost to the environment as waste heat.

These numbers suggested to Ackermann that the capacity of the absorber and the storage tank were large enough to permit continuous operation of a 1000-horsepower engine even with large heat losses (and they were expected to be enormous) and with extended periods of heavy cloud cover, or both. A substantial margin of safety was built into the design.

Shuman was particularly proud of the heat storage system—how the hot water was kept separate from the cool water returned from the turbine and how such a simple device solved the problem of providing continuous energy to the steam engine. He later wrote in *Scientific American*:

---

[2]This is the amount of coal one would need to supply the energy directly used by the steam engine. In practice, the boilers of coal-fired steam engines were so wasteful that it took about 15 tons of coal a day to run a conventional 1000-horsepower steam turbine. Most of this heat was never converted into mechanical energy; it simply disappeared up the flue as waste heat.

> *The hot water produced during the day is run into a properly insulated iron storage tank. Hot water is drawn from the top of the tank. After passing through the turbine, where its heat is transformed into power, it is pumped back through the bottom of the tank, displacing the hot water towards the top and causing it to overflow into the turbine. It requires very little power to keep up this circulation as it is almost entirely effected by gravity.*
>
> *To effect a proper upper displacement of the hot water by the cold water flowing in at the bottom, without mixing, the cold water must flow in and the hot water flow out in such a manner as to make the circulation very slow and evenly divided throughout the area of the tank. This we do by running the cold water in through a pipe at the bottom of the tank under a perforated partition, the water distributing itself evenly under the partition, and flowing through the small holes into the tank above, causing a perfectly even displacement of the hot water towards the top.*
>
> *At the top of the tank is a similar perforated partition about one foot below water level. From over this partition the hot water flows into the turbine. This has been thoroughly tried and works perfectly.*

The two partitions—colanderlike sieves—ensured that the hottest water would always be available at the top of the tank, where it could be taken into the steam engine, and the coldest water would always be at the bottom, where it could be drawn off during sunny hours to be heated in the 4-acre absorber. The storage tank, which had no moving parts and required no external energy source to keep the hot water separated from the cold, was an example of Shuman's inventive genius at its best—intuitive, inexpensive, practical, and effective.

The storage tank was to be buried deep below ground. Surrounding it with earth provided cheap insulation, keeping its iron walls away from heat-robbing air currents. Just as important, lowering the tank eliminated the need to pump water 30 feet above ground level. Because hot water rises (it is less dense than cold water) and cold water sinks, natural circulation alone would tend to bring the hottest water to the top of the storage tank. By keeping the hot water just at the level of the steam engine itself, it could flow into the engine without mechanical aid. By allowing the cooler water to sink to the bottom of the tank 25 or 30 feet below ground level, Shuman hoped again to reduce the mechanical energy that would be taken from the steam engine to run pumps (though he recognized that some pumping would be required to keep the water moving through the system in the volumes needed).

Steam at Low Pessure

A feature of the solar power plant that was likely to prove puzzling to potential investors with no background in mechanical engineering was the operation of the low-pressure turbine. It was easy enough to explain to a layman how a conventional turbine worked: Superheated steam under pressure in a boiler was released into the turbine, where it pushed against the blades of the turbine rotor, causing the shaft to rotate. This rotation was transmitted to gears, which connected the turbine to machinery such as a dynamo. A child blowing on a pinwheel illustrated the basic principle.

But where was the high-pressure steam to come from in a solar plant? The water was heated only to just under the boiling point at atmospheric pressure; at most, it gave off a few wisps of steam, like a cup of hot tea. How was merely hot water, under no pressure, to drive the blades of a turbine?

The solution depended on another physical property of water. Water's boiling point falls as air pressure is reduced. In a near-perfect vacuum, water will boil below 40 degrees Fahrenheit. In the perfect vacuum of outer space, liquid water cannot exist at all; it will boil at any temperature above its freezing point. By lowering the air pressure in the turbine to a near-vacuum, the hot water, heated to 212 degrees Fahrenheit, would turn into steam with tremendous force, driving the blades.

The 1000-horsepower steam engine for the proposed plant was one of the many turbine designs that had preoccupied Shuman in Philadelphia the previous year. It operated in a partial vacuum of

28 inches of mercury.[3] At that pressure, the boiling point of water drops precipitously: It falls from 212 degrees to 102 degrees Fahrenheit. The water would come into the partial vacuum inside the turbine containing "110 degrees more heat than it can hold," as Ackermann put it. The result could only be described as a controlled explosion. About 10 percent of the water would instantly be converted into steam, expanding 1600 times in volume. This force would be transmitted to the turbine rotor blades, causing them to spin. Ackermann described it this way:

> *The water which has not been exploded into steam is projected forward by the expansion of the steam, and the steam and water strike the turbine as a homogenous mixture, having velocity due to its expansion. In this manner, we are able to secure the full thermal efficiency of a drop in temperature [of 110 degrees].*

Thus, by creating a partial vacuum in the turbine, Shuman was able to secure from ordinary hot water some of the advantages of using high-pressure steam in an ordinary turbine.

The cooling steam emerging from the turbine would be collected in the condenser, also under near-vacuum conditions. From the condenser, the water would be pumped into the solar absorber, once again at atmospheric pressure, to be reheated to 212 degrees, or sent back into the storage tank to be stored for later reheating.

Those were the major features of the solar-powered plant. A small gasoline engine served as a starter motor for the turbine. It primed the air pump that maintained the vacuum within. It also started the water pumps that circulated water in the system and sucked water out of the low-pressure condenser. All these mechani-

---

[3]The strength of a vacuum, like atmospheric pressure, is measured in inches of mercury. A vacuum of 28 inches of mercury at sea level contains roughly the air pressure that would prevail at 50,000 feet above sea level. A "perfect" vacuum would be one measuring 30 inches of mercury, which is the full weight of the atmosphere at sea level. For comparison, the weak vacuum in a can of coffee before you break the seal is about 10 inches of mercury.

cal functions were taken over by the turbine itself once it was in full operation, and the gasoline engine was turned off. It was then needed only to restart the system after a complete shutdown.

Within days of Shuman's arrival in London, he and Ackermann were of one mind about the solar plant they would promote to potential backers. They ran the proposal by Ackermann's partner in his consulting engineering practice, C. T. Walrond. To take him through the plan, Ackermann prepared an hour-by-hour description of how the solar power plant would function:

> *When the sun rises, the heat absorber is already filled to its proper level with water used in the turbine the day before, which is at a temperature of about 100 degrees Fahrenheit, owing to loss by cooling overnight. Between sunrise and about 9 o'clock, this water is heated by the sun's rays, none of it being drawn off. Until 9 o'clock, the turbine has been run by water from the storage tank. At 9 o'clock, the storage tank is shut off, and the water from the south side of the sun heat absorber is circulated through the turbine at a rate sufficient to liberate enough heat to run it at the proper power. From the bottom of the condenser, it flows again into the north end of the sun heat absorber.*
>
> *By 11 o'clock, the water will have increasingly absorbed greater amounts of sun heat, and there will be a great deal more heat absorbed than the turbine can use. A gradually increasing portion of this hot water is run into the top of the storage tank and an equal amount of cold water run out of the bottom. The hot water, which the sun heats between 11 and 3 o'clock, is therefore used to run the turbine and also to fill the storage tank.*
>
> *About 3 o'clock, the storage tank being filled, the hot water from the sun heat absorber goes again direct to the turbine; flowing through that at a gradually increasing rate until about 5 or 6 o'clock. It is possible to run until about 6 o'clock with the water from the sun heat absorber, as there is a great deal of heat stored in it. This is gradually replaced by the cold water from the turbine. After 6 o'clock, the water from the storage tank runs*

the turbine throughout the night hours, the same cycle to start again the next morning.

Walrond enthusiastically endorsed the solar power plant proposal. Shuman and Ackermann declared that they were now ready to hit the speaking circuit to reach an interested public, some members of whom might lend financial support to the effort.

Making the Pitch

In the days before radio and television, public lectures and after-dinner speeches at scientific societies, civic organizations, and university-affiliated groups were (besides printed newspapers and journals) the most important mechanism for disseminating scientific information in Europe and America. Lectures were nowhere more popular than in Great Britain, which had a long tradition of scientific societies dating back to the prestigious Royal Society of London for the Promotion of Natural Knowledge, the oldest scientific society in the country and one of the oldest in Europe. The Royal Society, founded in 1660, was itself an offshoot of earlier groups that met informally to discuss scientific subjects.

By 1911 hundreds of lesser groups, some with pedigrees almost as distinguished as the Royal Society, were fixtures in every large and medium-sized city in England. They were organized mainly to discuss the two great subjects that had seized the public imagination since the middle of the previous century: the continued, spectacular advances in science and engineering and the exploration of remote parts of the globe, especially Africa.

Through his personal memberships in clubs and societies and the far-flung network of powerful Britons he had cultivated over 20 years as a consulting engineer, Ackermann had access to many of these groups. He took advantage of his connections to obtain speaking invitations for his American collaborator. Within a few weeks, word of Shuman's remarkable abilities as a public speaker had generated a second wave of invitations to still more.

The last months of 1911 were frenetic. Shuman's social schedule was packed with lunches and dinners. He seemingly spent all his time drumming up interest for his project by giving speeches to scientific groups, holding a private exhibition of his machines at Shaftesbury Pavilion, and attending dinner parties where he hobnobbed with the cream of England's financial and scientific community.

By all accounts, Frank Shuman was a terrific orator. A practiced dinner guest for many years at Philadelphia's most elite clubs and private homes, he was by 1911 at the top of his form, radiating charm, grace, and good humor as well as a wealth of solar information. As he now approached his fiftieth year, he cut a compact, vigorous figure, often compared to America's most popular recent president, Teddy Roosevelt, whom he superficially resembled. He had the gravitas and the curriculum vitae needed to command the attention of bankers and press barons. He soon discovered that as a homegrown American can-do engineer with little formal education but lots of inventive genius (in the mold of the ever-popular Thomas Edison), he was often treated as a matinee idol by London society. Shuman thrived on the public acclaim and used his bully pulpit shamelessly to advance his business venture.

Ten years of preoccupation with solar experiments in Philadelphia had cloaked Shuman's expertise in an almost messianic fervor. He believed in solar energy as a boon to humankind. He loved to begin his talks by predicting that one day massive sun collectors spread over miles of desert—in Arizona, the Sahara, and elsewhere—would provide almost all the world's energy needs. Power equal to all the coal and oil produced throughout the planet in 1911 (which Shuman had calculated to be some 270 million horsepower operating 24 hours a day the year around) could be produced, Shuman argued, from a square collector with sides 143 miles long covered with solar absorbers. Such a collector, 20,250 square miles in area, would cost $98 billion by Shuman's reckoning, an investment that would be fully recouped as the world exhausted its fossil fuels.

Given suitable appliances, Shuman averred, power of any desired amount could be obtained from the sun. Of the necessity for the use of this power and of the scope in the world for future sun power plants, there could be no question among the clear thinking. Shuman told his audiences:

> *Having planted the desert with sun heat absorbers at a cost of $98 odd billion, let us look at what we have secured at this enormous cost, and let us see whether we have made a good investment.*
>
> *We would have a plant that is worth to us at least as much as all the coal and oil fields in the whole world, because it can perpetually give us as much heat and power as all of the coal fields and oil fields of the world put together, if mined at the current rate. And these are certainly worth very much more than $98 billion.*
>
> *This vast investment would not be made for or by the individual, but for and by the entire human race, and we may safely assume the human race to survive all the coal and oil fields by many thousands of years. Hence, its overwhelming value of perpetuity and its capacity for practically infinite expansion.*
>
> *To the individual, $98 billion is a staggering sum, but to the human race, particularly if spread out through a period of say 200 years, it is almost nothing. The human race has expended in coal mines and oil mines and boiler and heating plants many times that sum during the last century alone.*
>
> *I feel sure that the greatest developments in sun power will come when the minds of many thousands of thinkers will be turned in this direction by the results of our work. We do not expect to do it all alone. All we shall now do is to establish sun power as a commercial rival of coal in those portions of the true tropics where coal is very expensive and the sun is very powerful.*
>
> *One thing I feel sure of and that is that the human race must finally utilize sun power or revert to barbarism, and I would recommend all far-sighted engineers and inventors to work in this direction to their own profit, and the eternal welfare of the human race.*

Though his preaching might have seemed naive to some, even hokey, Shuman had taken the correct measure of the enthusiasm of his age for grand, transforming, visionary projects, and his speeches were often met with standing ovations, even among jaded Londoners. Within weeks of the start of his speaking tour in England, Shuman could announce triumphantly that "sufficient money is at hand to go into business on a large scale, and there will be great developments in the near future."

Siting the First Plant

As subscriptions to the Sun Power Company (Eastern Hemisphere) Ltd. began arriving at Shuman's London bank, Shuman and Ackermann shifted their focus from the design of the sun plant itself, which they now believed pretty much settled, to its ideal location. Both men thought they could run a solar plant for less than the cost of a comparably sized coal-fired plant. Taking the increased investment for the solar absorber into account, Shuman believed they could compete with coal selling for as little as $2.50 a ton in those zones of the tropics where there was continuous sunshine.

To hedge their bets, Ackermann felt that, for the next decade or so, Sun Power Company should select only those locations where coal sold for $15 a ton or more and where they could count on sunny days 90 percent of the year or better. This would provide a wide margin of financial safety. Under such conditions, there was little doubt that the solar-power plants would have no trouble demonstrating they were cheaper than other alternatives, and would do so with a sufficient profit to accommodate unforeseen problems and costs.

Shuman had not much concerned himself until now with the actual work that his solar steam engines would perform. In the demonstration project in Philadelphia, he had connected the steam engine to a water pump that circulated and recirculated the same water 33 feet into the air, creating a visually striking but otherwise useless waterfall. The point was merely to show that the steam engine could do work. The waterfall impressed visitors. In a commercial setting, this would no longer suffice.

To be successful, not only was it necessary that the solar plant function as advertised, it was necessary that it do useful work—run a dynamo, power a mill, saw lumber, or operate machinery. What useful work could a solar plant do in a remote, dry location that was sufficiently low-tech not to imperil the success of the project nor distract attention from the solar plant itself? The answer the two men settled on was irrigation. They would use the solar plant to pump water into fields.

There was also a political dimension to the selection of a demonstration site. It would not do merely to have a great deal of sunshine; a clean, dry atmosphere; cheap land; and expensive coal—though these were all essential factors. Ideally, the demonstration plant should be located in a place where British policies were the focus of popular attention back home, guaranteeing press coverage for any new developments in the region. Better still, the location should be a place where the addition of a new source of cheap energy could spell the difference between the success and failure of British policies. Did any British colony fill that bill?

Such a place was soon found. As England's most fractious possession, its name was on the front pages of the London papers most days of 1910 and 1911. It was a territory that had caused endless troubles for the empire for nearly 40 years, creating burdensome expenses for His Majesty's chancellor of the exchequer, yet it was a place that, after the British Isles themselves and India, was the most important strategic center for the exercise of British power in the world in the years before World War I. It was also a place that served as guardian of the vital sea lane to India, where a succession of British prime ministers had painstakingly sought to consolidate power from 1875 on, and where coal was outrageously expensive. Such a place was Egypt.

Britain's permanent entanglement in Egypt dated back to the construction of the Suez Canal, which was begun in 1859 and completed after many difficulties 10 years later. This vast, visionary engineering enterprise had put Egypt deeply into debt. It also put Egypt on the world map. By greatly shortening the distance by sea from Europe to Asia and the Persian Gulf (by making it unnecessary to circumnavigate Africa), the canal immediately raised Egypt's strategic importance in the eyes of all the big nations of Europe. In particular, it shifted Great Britain's focus in the Middle East from

Constantinople (the seat of the old Ottoman Empire) to Cairo, and it opened the door to British intervention in Egyptian affairs.

Unfortunately, the problem of the unpaid canal debt and its mounting interest was compounded by the prodigality of Egypt's khedive, Ismail Pasha, whose fondness for building personal residences, palaces, colossal public buildings, irrigation projects, and schools all but guaranteed that a bankrupt Egypt would soon fall under British control. Ismail outraged conservative British politicians by running up Egypt's national debt from 3 million to 100 million pounds sterling in 11 years, from 1864 to 1875.

With Egypt on the brink of bankruptcy, the khedive's European bankers refused to extend him credit. Faced with crippling debts and a devastated economy, he resorted to a solution that ensured Egypt's subjugation to the British Empire for the next half century: on November 25, 1875, the khedive sold Egypt's stake in the Suez Canal Company to the British government for 4 million pounds sterling.

The British government thus became the largest single shareholder in the canal, so Britain had to ensure the security of Suez. No other nation could be permitted to occupy Egypt or the canal without threatening the sea-lane to India and England's imperial greatness. Unfortunately, Egypt's financial situation continued to worsen. Khedive Ismail squandered his 4-million-pound windfall in less than 12 months, and in April 1876 his hapless finance minister suspended payments on all Egyptian bonds. Ismail was forced to accept a French-British "debt commission" to manage Egypt's financial affairs and, effectively, to run her economy. This gave rise to a smoldering resentment of foreign control that would dog the British occupation until Egyptian independence half a century later.

Egypt's financial problems were soon aggravated by political chaos induced by the outbreak of a nationalist revolt in 1881. The Egyptian army mutinied. The country was controlled by mutineers and remained in a state of panic for almost a year. The following June, 50 Europeans in Alexandria were slaughtered "under circumstances of the utmost brutality," in the words of the British consul of the time, who was himself severely wounded in the attack.

The British reacted with the naval bombardment of Alexandria in July 1882 and by landing a British expeditionary force at Ismailia. On

September 13, 1882, British forces prevailed. Prime Minister Asquith was determined not to let matters in Egypt get out of hand again. He immediately sent to Cairo Britain's preeminent diplomat, Lord Dufferin, who confirmed that the British had inherited a country in shambles. Cholera had wiped out more than 100,000 Egyptians in the previous 12 months. Egypt's cities had no sewers. Sanitary conditions in the towns and villages were "frightful, and too revolting to write about," in the words of one reformer. There were no hospitals or physicians.

Recognizing how much of a drain on the British treasury a permanent occupation of Egypt would represent, the British government shrank from taking on that obligation, but the exposure of the Suez Canal, now British property, permitted no alternative. In 1883, the British consolidated control over Egypt. Over the next 30 years, the country was effectively ruled by British administrators, first by Sir Evelyn Baring, whose 24-year administration effectively made Egypt and the canal a part of the empire. In 1907, Sir John Eldon Gorst succeeded Baring as British agent and consul-general. A rising tide of nationalism marred his term, and the British grip on Egypt appeared to be loosening dangerously once again. Besieged by criticism, in 1911 Gorst returned to England.

Prime Minister Asquith was determined to replace Gorst with someone who would put an end to domestic concerns about Egypt, to demonstrate unequivocally that Egypt was firmly under British control. He settled on the most famous military man of his era, Field Marshal Sir Horatio Herbert Kitchener, a soldier brimming with titles and decorations and a popular national hero in England from the time he had led the Anglo-Egyptian reconquest of Sudan a quarter-century earlier.

A few weeks before Frank Shuman and A. S. E. Ackermann settled the question of a suitable location for their first solar plant, London newspapers announced that Lord Kitchener of Khartoum had accepted the government's nomination as His Majesty's agent, consul-general, and minister plenipotentiary in Egypt. Public sentiment was overwhelmingly in favor of the appointment. Egypt's problems—financial, political, and social—had loomed too large in the public eye, and Lord Kitchener was just the man to restore confidence that Egypt could be made to toe the British line.

General Horatio Herbert Kitchener, later Lord Kitchener of Khartoum, around 1896 when he was in his mid-forties and sirdar (commander-in-chief) of the Anglo-Egyptian army. Ladies in Cairo often swooned when he entered a room. He would become one of Frank Shuman's biggest supporters.

Kitchener was not a man to ask opinions about the role he should play in Egypt—he gave them. His style always had been to announce his goals and knock down anything or anyone who stood in the way. He recognized the central importance of coming to terms with rising Egyptian nationalism. With the paternalism that was characteristic of British imperial policy, Kitchener firmly believed that, after a period of instruction in the theory of British parliamentary democracy and an infusion of industrial capital, any country could be trained by Britain's colonial administrators to run its own affairs harmoniously. This was an article of faith in nineteenth-century Whitehall and would remain so until well after the First World War.

To prepare the way for an Egypt that would slowly become more independent, Kitchener announced before leaving England that he would put emphasis on improving Egyptian infrastructure. He wanted to build roads, raise dams, increase Egypt's arable land through irrigation, double her cotton production—he had a long list of goals. He talked about his plans extensively in the British press. Agriculture and irrigation were major themes. "Prosperity and water go hand in hand in Egypt," Kitchener announced before boarding a steamer for Africa.

To accomplish these feats, Ackermann and Shuman realized, Kitchener would need energy, and vast amounts of it—but it was energy that Egypt did not possess. Coal, imported from Britain and Germany, was selling in Egypt's interior at a peak price often above $40 a ton, more than 15 times its cost at the mine mouth. All Egypt's coal was imported. The nation's wood supply had disappeared millennia earlier, in the time of the pharaohs, and the country's peasants, the fellahin, now burned camel dung for fuel. Though the smallest of the three dams that would eventually be built at Aswan had been raised in 1902, hydropower would not come to Egypt for many years—not until the third dam, the Aswan High Dam, was built in the 1960s.

After a well-publicized round of dinners and speeches elaborating the great undertaking before him, Kitchener set sail for Alexandria and his proconsulship on September 20, 1911. "Lord Kitchener conquered Egypt before he sailed from England," wrote British journalist Sydney Moseley, who went with him. "The unanimous acclamation from the British press will always be remembered as having been the prelude to Lord Kitchener's success as British Agent in Egypt."

Shuman and Ackermann knew a good risk when they saw one. They judged Egypt under Viscount Kitchener of Khartoum the ideal place in 1912 to put solar steam power on a commercial footing. The two engineers put into motion their own plans for Shuman's speedy departure for Cairo and the flood plain of the Nile. It was an enterprise that would engage both men for the next several years.

As it happened, Shuman's mission was remarkably similar to that of an earlier solar engineer who had also set out from London to bring solar power to a distant British possession in the tropics. Had he thought about that adventure and its outcome, for he surely had heard of William Adams and his experiments in India, Frank Shuman's confidence en route to Cairo might not have been as great as it was.

4
"A Substitute for Fuel in Tropical Countries"

The Sun Power Company's foray into the upper delta of the Nile in 1911 and 1912 was not the first attempt to harness solar power in a tropical possession of the British Empire. That effort had taken place more than 30 years earlier in India, guided by the only Englishman before A. S. E. Ackermann to work on practical applications of solar energy. He was William Adams, an engineer who served as deputy registrar of the High Court at Bombay.

Before leaving Great Britain for his colonial post in 1870, Adams had served in the British patent office in London. His work with patents led him to become something of an amateur historian of energy technology. Windmills and waterwheels fascinated him. He saw the history of England through the prism of its energy use, the story of a river of energy transforming the nation with an astonishing speed.

The evolution in Britain's use of energy—from human labor to animal muscle power to windmills and watermills and later to steam engines—was slow at first, spreading over the centuries. The pace picked up in the eighteenth century. By the middle of the nineteenth century, the changes wrought by industrialization, accelerated by the power of the steam engine, were unfolding in years and decades rather than centuries and millennia.

Unlike his contemporaries, who saw the steam engine as a wonderful but isolated discontinuity—a paradigm shift in England's energy development—Adams located the nineteenth century's Age of Steam in its historical context as part of a broader evolution. Steam engines were not the first machines to harness energy for human work. Early British waterwheels, those made before 1000, were more than 6 feet in diameter and could generate 3 to 4 horsepower, easily the most powerful prime movers[1] of their time. They powered the earliest British machines, which were mills to grind grain.

Waterwheels were later adapted to drive sawmills and pumps, to provide the bellows action for furnaces and forges, to drive trip hammers for working red-hot iron, and to provide direct mechanical power for textile mills. Until the development of steam power, waterwheels were the main source of mechanical power in the British Isles, rivaled occasionally by windmills.

The Domesday Book census of 1086 lists 5624 mills in some 3000 English villages south of the Severn and Trent Rivers, many powered by water—more than 1 mill for every 400 Britons. By the eighteenth century, mill workers had produced an ingeniously sophisticated technology for their waterwheels with intricate linkages—gears and chains and drive shafts—to transmit power *3 and 4 miles* from the wheel to the sites where the power was needed to drive machines.

Waterwheels were found throughout Europe, where they were used even more broadly for labor-saving tasks than in England, from pressing the oil from olives in Portugal to rolling and pulling wire in Germany. Some ancient installations were surprisingly large. The Romans built a mill in the fourth century with 16 wheels, each more than 7 feet in diameter, at Barbegal on the Arcoule River. The flour production of the mill neared 3 tons a day, far more than could be consumed locally. The surplus was shipped back to Rome from the nearby port of Arles.

---

[1]To all but the theologian, a "prime mover" is an initial source of motive power—whether a waterwheel, windmill, steam engine, diesel engine, or turbine powered by a nuclear reactor.

The "Lady Isabella" waterwheel at the Laxey mine on the Isle of Man.

The most spectacular modern example of waterwheel technology in Adams's day was the "Lady Isabella" wheel at the Great Laxey Mining Company on the eastern shores of the Isle of Man. Overlooking the Irish Sea beneath the rugged, windswept hills of Snaefell, the loftiest Manx peak, Laxey employed 600 miners to quarry zinc blende from the unyielding rock half a mile below the surface. Laxey was by far the largest mine in the British Isles. Besides zinc, it produced smaller quantities of lead, silver, and copper ore, all shipped to England and Wales from Laxey's modest harbor.

The giant wheel was built to drain the mine in 1854, long after the introduction of steam power in England's mining industry. It was 72 feet in diameter (the height of an 8-story building), and 6 feet wide, and it could deliver between 250 and 400 horsepower, depending on how much water—snowmelt from Snaefell's slopes 2000 feet above the village—was let through the iron sluice at the top of the wheel. The shaft was turned by the weight of water in 168 buckets, each with a capacity of 24 gallons. With Lady Isabella's power, the mine operators could lift 15,000 gallons of groundwater an hour from a depth of 1500 feet in the mine.

Windmills were a more recent but still ancient technology, dating at least to the tenth century in the Middle East, a bit later in Europe. In one form or another, windmills remained in use for milling grain, pumping water, working metal, sawing wood, and crushing chalk.

Though Britain's transformation into a society that used prodigious quantities of energy was associated with the development of that revolutionary new prime mover, the steam engine, Adams recognized that before this means of energy conversion became available the efficiency of windmills and waterwheels had already improved dramatically. The Lady Isabella was hardly the only example of modern interest in these technologies. The nineteenth-century British windmill, with its easily controlled spring sails, self-regulating fantail, and centrifugal governor, was as much a product of advanced industrialization as the steam engine itself. Even without the steam engine, British industrialization would likely have made great strides in the nineteenth century through wind and water power (both continued to play an important role in industry throughout Adams's lifetime), though the scale of the transformation would have been smaller.

In Adams's youth, steam engines were already used to pump water out of mines. James Watt's improved models could power any machinery. By the mid-1800s Adams had calculated that steam power was doing the work of 2.5 million Britons at a time when the census had identified about 3 million people engaged in manufacturing overall in the nation. That meant that by 1850 steam machinery was already performing work equivalent nearly to that of the *entire* factory work force of Great Britain. It was only a matter of time, he thought, before steam-driven machines doubled, tripled, or quadrupled the output of the human work force in England, with predictably huge effect on an already strong economy.

The First Solar-Powered Steam Engine

Adams devoured technical reports about energy and engineering, and his interest in wind, water, and steam power led him naturally to solar energy. He recounts his excitement at discovering an article in the French scientific journal, *Revue des Deux Mondes*, about Augustin Mouchot's solar steam engine in Paris (discussed in detail in Chapter 6). That machine, which provided power for a printing press, is generally regarded today as the first solar-powered machine that was not a toy. It was exhibited at a Parisian world's fair for

Napoleon III, where it was billed as "the first solar steam engine that ever worked in Europe."

What impressed Adams about Mouchot was the practical bent of his work. Not for Mouchot was research "only for scientific curiosity" that led to "erroneous conclusions drawn from unsound premises which would have been avoided by practical men."

In his introduction to *Solar Heat: A Substitute for Fuel in Tropical Countries*, the book he published in Bombay in 1878, Adams lamented that European researchers of his time took only an academic interest in solar power. There were only a few, he wrote, who

> *have maintained that the rays of the sun could be used and made to perform most purposes for which ordinary fuel is used, especially for driving steam machinery. One of these, M[onsieur] Mouchot, Professor of Mathematics in the Lycee of Tours, in France, has proved the possibility of this by actually setting a steam engine in motion, entirely by concentrated solar rays; and he and Professor Rohlf, the celebrated African traveler, and some French savants, are said to be making experiments with the object of using solar heat, concentrated by gigantic revolving burning glasses, as a motive power for railway locomotives on a line that they propose to construct across the desert of the Sahara. This idea may be, and probably is, purely Utopian, but very important discoveries have been made in striving for the impossible; and if no further success is achieved than that of utilizing the rays of the sun for driving stationary steam engines, an important addition to physical science will have been made, and a great commercial revolution will have been effected.*

We hear no more of the celebrated Professor Rohlf or of his (certainly imaginary) locomotive, but the seed had been planted in Adams's mind. He believed he had found just the right niche for solar power. "The standard works published upon heat make very little mention of the concentration of solar heat, and they are silent on the question of the possibility of using the rays of the sun as a substitute for fuel in tropical countries." These words, the first sentence in Adams's little book, could just as easily have been written by Frank Shuman. In India, blessed with bountiful sunlight, Adams was eager to investigate practical ideas that could be used locally. He wrote:

> *The results of experiments show such strong grounds for believing that solar heat can be extensively utilized that it is a matter of astonishment that, until now, the idea has never been submitted to the test of experiment on a large scale in India, where there is by no means a lack of solar heat to concentrate.*

The purpose of his book, he tells us, "with a view of filling up this blank in research," is to record the results of "a series of experiments in Bombay, the results of which convince me completely that solar heat can be extensively used."

Steam engines, like human laborers, had rapacious appetites. In England, steam engines were fueled by coal because British forests had been depleted of cheap wood for burning. Those hardwoods that remained were needed to build ships, not to waste as fuel. Though Adams had succeeded in patenting a crude solar boiler in 1860, the industrial revolution's commitment to coal was already too powerful to be resisted in England. British engineers derided the notion of solar energy as a substitute for coal. If coal mining was dangerous, then it was only a matter of bringing in cheap (and expendable) Irish and Welsh labor to man the pits and tunnels. Adams's solar ideas were rebuffed in his homeland. Britain was a dark and cloudy place and too far north to benefit from solar energy.

Taming the Bombay Sun

Like researchers before him and since, Adams fell under the spell of Athanasius Kircher. Adams's first experiments in Bombay were reenactments of those of the great Jesuit alchemist, though his version of "Kircher's mirrors" had several new twists unknown to Kircher. First, Adams had a local carpenter build a set of four 3-tiered

One of a set of four 3-tiered wooden shelves William Adams built to align arrays of plane mirrors in his duplication of "Kircher's mirrors."

wooden shelves that could each hold 18 "looking glasses." These mirrors, 17 by 10½ inches, were set in wooden frames with a lever in the back enabling him to adjust each to reflect light on a single spot. Adams describes how he designed his mirrors:

> *Each glass was moveable on a swivel in the same manner as an ordinary toilet glass, and also upon a pivot driven through the foot of the frame, so that it had a universal motion by which the reflection could be directed on any object, and to any distance, by the touch of the finger. I had four sections of 18 each [constructed].*

Adams then describes how he put this apparatus to work:

> *Suspending a thermometer with a blackened bulb against a whitewashed wall, I reflected upon it the solar rays that came through a chamber window. I used, at first, one flat glass only, and I found that the mercury, which stood at 80 degrees Fahrenheit, rose to 102 degrees in ten minutes. I then added the reflection from a second glass and the thermometer rose to 124*

degrees in ten minutes. A third reflection in less than ten minutes raised the mercury to 140 degrees, beyond which the thermometer was not built to register.

Simple calculation showed that each mirror had raised the temperature 22 degrees. The heat rise confirmed for Adams, as it had for Kircher, that plane mirrors were just as effective in concentrating solar heat as more expensive lenses and concave mirrors.

Taking his apparatus outdoors, he found that

in the open air and with the reflection at a greater distance—20 feet—the increase per each reflection was considerably less than 22 degrees Fahrenheit. On trying 18 glasses upon another thermometer that professed to register up to 700 degrees Fahrenheit, I only obtained 360 degrees. Deducting 90 degrees as the initial temperature (for the thermometer was shaded from the direct rays of the sun), this gave me an average of 15 degrees for each reflection.

When he deployed two of his banks (holding 36 mirrors), he found that he could "make the mercury in the thermometer really

Two banks of the giant compound concave mirror Adams built to heat a 12-gallon boiler.

boil, leaping up to over 670 degrees Fahrenheit." With all four of his units in use, the temperature rose beyond the ability of his best thermometer to record it, although he estimated that it was around 1140 degrees.

"A copper cylinder containing three gallons of water, placed in the focus, boiled in exactly twenty minutes," he recounted. "Wood ignited immediately."

Adams ordered his Indian *fundhi* (skilled carpenter) to build more banks of mirrors, this time fixed in their wooden frames to create a giant compound concave mirror shaped like a portion of the inner surface of a sphere with a 24-foot diameter, so that the reflections from all the flat mirrors, when the wooden shelves were properly arranged in a semicircle, intersected at a distance of 12 feet. Each unit of 20 mirrors could independently generate a temperature of 300 degrees Fahrenheit at a focus about a foot in diameter. His success with this solar furnace, and his written account of it, won the gold medal of the Sassoon Institute of Bombay for the best new and useful invention.

Adams felt he was ready to move beyond simple experiments. He decided to approach the problem of solar collection solely in terms of concentration of the solar flux, making unnecessary the need to tinker with the steam engines at the receiving end of the power. He was convinced that inexpensive flat mirrors made the job of concentrating solar heat economic. This would allow his solar-powered boilers to achieve the high temperatures and pressures needed to power conventional steam engines.

India's New Energy Source

The Bombay of William Adams's day was a teeming city of street vendors, shops, and stalls. Though he worked in a government building near the central fort, Adams's outdoor laboratory was on a spit of land called Colaba near Back Bay and the open sea. There he could get clear access to the sun from dawn to dusk, beyond the reach of buildings and the shadows they cast. A warren of alleys led south from the fort to the pungent Sassoon Dock, where fishing boats and Arab dhows arrived to unload cargo. The fort was dominated by Victorian Gothic buildings that sent a pleasant shiver of recognition down the spines of the *pukka sahibs* from the industrial cities of northern England, men who now sat at the top of the city's social and business hierarchy. The area around Sassoon Dock, in contrast, was the "native quarter"—and a kaleidoscope of humanity and world commerce.

India was as wood-poor as England and desperately needed new sources of energy. In fertile areas like Bombay that were heavily populated, humans quickly outstripped plants' ability to store the sun's energy. Photosynthesis is the process by which plants convert sunlight, carbon dioxide, and water to "biomass"—that accumulation of complex organic compounds including sugars, starch, cellulose, and oils that make up living plants and that can be oxidized to release heat. By comparing the amount of solar energy that reaches a plant with the amount that it converts into chemical bonds and stores within its cells, scientists in the nineteenth century showed that nature is not very efficient at storing sunlight. Only 0.1 to 2.0 percent of the energy that reaches a plant from the sun, they learned, is converted into biomass.

Some plants are better than others at manufacturing high-energy products. For example, sugarcane is at the upper end of the efficiency scale. In the nineteenth century, sugarcane could make 4 tons of sugar per acre per year, which is equivalent to 2 tons of ethyl alcohol or 1.2 tons of ethylene. In contrast, trees are least efficient at converting sunlight into stored energy. The heat released by a log burning in a fireplace is a minuscule fraction of the solar heat it has taken to make it—the heat equivalent of a thousand burning logs, enough to keep a house warm all winter. Such is nature's prodigality.

Bombay's hot climate, its moisture-laden air coming off the Arabian Sea, is ideal for cultivation. But the Bombay of the last half of the nineteenth century was growing so rapidly in population that fuel grown from plants, especially firewood, had been depleted for miles around the sprawling island city. Even underbrush was scarce.

There were few other natural-energy options. Surrounded by flat wetlands stretching boundlessly into the subcontinent, Bombay lacked the changes in elevation that make damming water possible.

The Arabian Sea dampened temperature changes. In the warmest month of the year, May, the temperature hovered around 90 degrees Fahrenheit. In January, the coolest month, it rarely fell below 75 degrees. The monsoons between June and September dropped 70 inches of rainfall on the city annually, with the showers arriving like clockwork on the seventh of June, year after year. Bombay's hot, moist air hung over the city like a pall for months, motionless, interrupted by violent monsoons that would have torn the vanes off any windmill. Water power and wind power were not possibilities for Bombay, though the region cried out for more energy.

And yet, whether energy was available or not, the British were determined that Bombay should grow as an important commercial center. Isolated from the rest of India, this growth was spurred in Adams's time by the arrival of British steamships and the construction of the first railway in Asia (initially with a mere 21 miles of track) from Bombay to Thana, a suburb. The Great Indian Peninsular Railway Company, in keeping with British tradition, introduced first-, second-, and third-class carriages for this short journey. The third-class coaches were exceptionally uncomfortable with no seats; the windows could be reached only by giants. The locals soon dubbed these coaches *bakra gadi*—goat carts.

Cotton mills appeared around the city in the 1860s. The American Civil War, which had exhausted Britain's supply of cotton, sparked a Bombay cotton boom just as it had in Egypt. By 1869 the opening of the Suez Canal and the massive expansion of Bombay's docks cemented the city's future as India's primary port. It was clear that Bombay's energy needs could no longer be met locally. The wealthy imported coal; the poor burned animal dung.

High-Pressure Steam from the Sun

Bombay's city fathers, open to innovation in this period of phenomenal growth, seemed receptive to Adams's solar experiments in a way London's political elite had not. Adams was interested in using concentrated sunlight to make high-pressure steam, the kind of steam that could run the most modern steam engines. To accomplish this, he realized he would need high-pressure boilers beyond the technical resources of Bombay's skilled metal workers.

"The boiler I used, I had made in England," Adams wrote. "It held 12 gallons. It was made of beaten copper ¼ inch thick, the strongest boiler in the world of its dimensions (2 feet 7 inches high by 18 inches in diameter)." Adams was aware that his French colleague Augustin Mouchot had worked with copper boilers only 3 millimeters thick, in which Mouchot generated steam pressures of 5 atmospheres (75 pounds per square inch). Mouchot's boilers were barely able to withstand that pressure, and he often had to shut down his equipment for fear of explosions. In *Solar Heat,* Adams wrote:

> *I ordered mine to be made ¼ inch thick and of beaten copper to the astonishment of the boilermaker, who in answer to my inquiry whether there was any fear of an explosion, said he "should think not!" Had he shown the slightest hesitation, I should have told him to make it ½ inch thick. Amateur engineers cannot be too cautious.*

When it arrived on a ship from London in 1876, Adams set up his copper boiler, painted black now to absorb the sun's rays, on a Back Bay beach in the focus of his giant compound mirror. At seven-thirty in the morning he turned the concentrated rays of the sun on the metal cylinder filled with water. Adams watched the equipment wearing dark glasses. He described his apparatus with a certain literary flair:

> *The luminous rays of the sun are transformed from light into heat by contact with the blackened shell of the boiler, so that every unit of solar heat is utilized by absorption into the metal, on which the rays beat like missiles in a continuous and incessant storm of solar fire, penetrating like the heat from a blowpipe, and the heat is thence communicated to the water. The solar heat thus accumulates in the metal as water accumulates in a lock. By 8:00* A.M. *there was a pressure of 10 pounds per square inch in the boiler; by 8:30* A.M.*, this had risen to 70 pounds.*

At 70 pounds per square inch of pressure, a safety valve installed by the boilermaker kicked in, releasing some of the steam. Adams's butler, who had come along to help set up the equipment, put a brick on the safety valve to keep the high-pressure steam from escaping. Pressure continued to build "when suddenly the packing and red lead at the top of the dome under the socket of the steam pipe gave way and, with a terrific noise, the whole volume of steam rushed out of the opening."

Fortunately no one was injured in the blowout. Relieved that he had barely averted a catastrophic explosion, Adams shut down the solar mirrors and opened up the boiler. What superheated water remained in the vessel rushed out in a cloud of hot gas, leaving the boiler completely dry. All 12 gallons of water had been vaporized. Adams explained how this near-disaster did not deter him from proceeding to the next step:

> *When this boiler had been properly fitted up again, a steam pump was hired, said to be of 2½ horsepower, and it was connected with the steampipe. At 7:30 A.M., the solar fire was opened on the boiler from the whole battery of 16 mirrors at a range of 20 feet, the boiler containing 12 gallons.*

An hour later, there were 55 pounds of pressure in the boiler, more than enough to operate the pump. Adams continued:

> *This pump, the first steam engine ever worked in India by solar heat, was kept going daily for a fortnight in the compound of my bungalow at Middle Colaba in Bombay, and the public was invited by a notification in the daily papers to witness the experiments.*

The British governor and other dignitaries dutifully trooped by to see the machine. Adams spent the next few months perfecting the apparatus, adding a circular rail around the boiler to permit the banks of mirrors to follow the sun as it crossed western India's blue skies. He redesigned elements of the boiler to increase the area of solar-heated metal in contact with the water, decreasing the time it took to develop a good head of steam.

After 12 months of additional tests, Adams was prepared to assert to the British authorities that "the results of my experiments with flat mirrors ascertain beyond the possibility of any doubt that stationary steam engines of any power can be driven at an expense of 2 pounds 10 shillings per horsepower in India on any day the sky is clear."

Without disputing Adams's results, Sir Richard Temple, Bombay's governor, after consulting a number of British engineers, concluded that solar heat "could not be used for commercial purposes on a large scale, such as a motive power for spinning and weaving mills." The problem for Temple was the familiar one of energy storage. The local mill workers arrived at work well before dawn, before a solar boiler could be fired up, and the factory owners were unlikely to agree to start their mills later in the day "or to give the workmen a holiday on days when the sky is not clear."

Adams scoffed at these objections, noting that any solar-produced steam would be an inexpensive supplement to that produced by burning coal. He asked:

> *Cannot steam generated by solar heat be used as an auxiliary to, and with, steam generated by fuel? Other engineers are unanimously of the opinion that it can. Steam generated by solar heat could be injected into the ordinary boilers, and every pound of steam so injected would represent a savings of precisely the quantity of coal that would have been needed to produce it.*

It was a persuasive argument. In a country where all coal was imported from England, where overland transportation costs of coal made its use prohibitive in districts far from the coast, and where wood "is always dear," Adams estimated that something like 25 percent of the fuel imported by the colonial administration could be saved by using solar-heated water to boost, rather than replace, coal-fired boilers.

Although some individual mill owners did build auxiliary solar heaters for their steam engines, most did not. Adams's frustration with the Bombay government soon led him to abandon further experiments, concentrating instead on writing his book.

As his research became more widely known, Adams got a more sympathetic hearing from another British enclave, that of Aden at the eastern mouth of the Red Sea, where potable water distilled from seawater "sold for the price of champagne." At Aden, the fiery hot, inhospitable port serving Yemen Protectorate, fresh water was scarce. Well water was brackish and freshwater streams were seasonal, turning into dry gullies in those months their water was most needed.

To maintain the colony, the British operated six coal-fired boilers and condensers to distill drinking water from seawater. These stills could make something like 9 to 12 pounds of fresh water per pound of coal burned—all the latter imported from England at great cost. Given these extreme circumstances, the British officers in command at Aden were willing to investigate *any* alternative to make potable water. They adapted Adams's apparatus successfully and used it to produce drinking water until cheap oil from the Arabian peninsula supplanted coal after the First World War.

In spite of the reverses he suffered, Adams remained committed to a conception of solar power, as evidenced in his book, different from that developed later by Shuman and Ackermann. He felt that cheap plane mirrors provided an economical solution to the problem of concentrating sunlight, enabling anyone to generate steam at the same high pressures produced by conventional boilers. This gave the advantage of being able to adapt solar boilers to all steam engines. Adams was not constrained, as Shuman later would be, to powering low-pressure machines. As Adams wrote:

> *With these facts and data, it is clear that solar heat can be generated by reflection from a combination of flat mirrors to a degree of intensity, and to an extent, that has never before been dreamt of. The heat of the fiercest blast furnace in an iron foundry may be described as genial warmth in comparison with*

the heat that can be generated by reflection of the solar rays from a combination of flat mirrors.

Adams published his book in 1878. There is today a fine presentation copy of *Solar Heat* in the noncirculating collection of Columbia University's Butler Library in New York, autographed by Adams and inscribed "with the author's compliments to Captain Ericsson, Bombay, 5th May, 1879." On page 14 of the printed text, Adams praises John Ericsson, "the engineer who invented the monitor turret ships, and who was the author of several other ingenious inventions—a Swede who became a naturalized American—who turned his attention toward the end of his life to the subject of the concentration of solar heat for driving steam machinery."

Free Fuel from the Sun

Like his contemporaries, John Ericsson feared that rapid consumption of coal in industrial economies would doom them because the coal, he believed, was already running out. In one of the many scientific papers on the commercial use of solar energy he published in the 1870s, he wrote:

I cannot omit adverting to the insignificance of the dynamic energy which the entire exhaustion of our coal fields would produce, compared with the incalculable amount of force at our command, if we avail ourselves of the concentrated heat of the solar rays. Already, Englishmen have estimated the near approach of the time when the supply of coal will end, although their mines, so to speak, have just been opened. A couple of thousand years dropped in the ocean of time will completely exhaust the coal fields of Europe, unless, in the meantime, the heat of the sun be employed.

It is true that the solar heat is often prevented from reaching the earth. On the other hand, the skillful engineer knows many ways of laying up a supply when the sky is clear and the great store-house is open, where the fuel may be obtained free of cost and transportation.

Swedish-born John Ericsson in 1839, aged 36, at the time he immigrated to the United States, where he became an American citizen 9 years later. Best remembered today as the naval architect who designed and built the Civil War ironclad *Monitor*, he was one of the earliest U.S.-based inventors to tackle the problem of solar-powered steam engines.

Ericsson shared with Adams a belief that solar-powered machines could contribute immediately to industrial growth by cutting fuel costs in the tropics. As early as 1868, in a paper Adams had surely read, Ericsson predicted that commercial use of solar power near the equator was just around the corner:

> *A great portion of our planet enjoys perpetual sunshine. The field therefore awaiting the application of the solar engine is almost beyond computation while the source of the power is boundless. Who can foresee what influence an inexhaustible motive power will exercise on civilization and the capability of the earth to supply the wants of the human race?*

Spurred in part by a purely scientific curiosity—he had once resolved, as he said, to measure for himself "the intensity of that big fire which is hot enough to work engines at a distance of 90,000,000 miles"—Ericsson would dedicate the last 20 years of his life (he died in 1889) to the creation of solar machines.

John Ericsson is best remembered in the United States as the designer, promoter, and builder of the Union's ironclad battleship *Monitor,* the vessel that defeated the Confederate ship CSS *Virginia* (known as the *Merrimack* by Union forces) at the Battle of Hampton Roads in March 1862. The *Virginia* was the Confederacy's secret weapon, a metal-sheathed battering ram capable of reducing wooden ships to flotsam. President Abraham Lincoln, with his personal interest in mechanical devices and a predisposition for talented inventors who might equip him with the latest and deadliest weapons of war, had ordered Union forces to build a steamer that could take on Confederate ironclads. Ericsson's was one of the first designs submitted for review, along with those of better-known naval architects.

Ericsson's plan created a furor. In an era still wedded to wooden vessels, the ship he proposed was to be made entirely of iron and most of the ship would be below the waterline. The crew's living quarters and the ship's engine would be submerged (a hallmark of ships designed by Ericsson—in the days before torpedoes, he thought sailors and engines were safer underwater, where they were less vulnerable to surface gunnery). In the middle of the deck, Ericsson proposed a 9-foot-high pillbox-shaped revolving turret that would hold two giant cannon. A little over 170 feet long, the *Monitor* would be half the length of the *Virginia,* small for a naval vessel. The U.S. Navy review board rejected the design as just too weird to succeed, and naval officers publicly ridiculed Ericsson's conception as "a cheese-box on a raft."

Undaunted, Ericsson sought a private meeting with Lincoln (perhaps the most accessible of presidents) to defend his proposal. Lincoln agreed to see him and became quickly convinced of the ship's merits. He awarded Ericsson the contract over the U.S. Navy's objections.

On January 30, 1862, the *Monitor* left her dry dock in Brooklyn and made her maiden voyage. Contrary to the predictions of her critics, the *Monitor* did not plunge to the bottom of the river but floated buoyantly. On March 6, her sea trials successful, the *Monitor* left New York Harbor with a crew of 57 sailors and headed south for blockade duty off the Confederate coast. Wholly steam-powered with a screw propeller, she reached Hampton Roads, the broad channel of brackish water through which the James, Elizabeth, and Nansemond Rivers flow into Chesapeake Bay, on March 8, 1862. The next day she engaged the apparently invincible Confederate ironclad ram—and made history.

After the victory, the *Monitor*'s detractors changed their tune. The design was now hailed as revolutionary and Ericsson's ship set a new standard for warships that continued well after the Civil War. Ericsson's new-found supporters argued that but for the *Monitor* the Confederacy might have dominated the strategic coastal waters off Virginia and the Carolinas far longer. Ericsson's ship was credited with having changed the course of the war. The Lincoln administration ordered many more *Monitor*-type vessels from Ericsson-controlled companies, making him a wealthy man in the years after Lee's surrender.

Though he was lionized after the Union victory, the opinions of others never seemed to carry much weight with Ericsson. He had been written off many times in his life and would continue to be, despite his fame. But the new money was welcome; now he could devote himself to building solar-powered machines.

Ericsson had been involved with much of the new technology of the mid-nineteenth century. After serving in the Swedish army as a topographical surveyor, he went to London to seek financial support for a new type of heat engine he had invented, which used the expansion of superheated air instead of steam as its source of power. While pursuing this venture, Ericsson formed a partnership with John Braithwaite, a rising star in British railroad engineering circles. In 1829 the two men produced a steam locomotive, the *Novelty*, and entered it into a national competition at Rainhill, Lancashire.

Steam locomotives at the time were considered an untested and untrustworthy technology. In the late 1820s the builders of the Liverpool & Manchester Railway were locked in a debate over whether to use horses or steam locomotives to pull their trains. They decided to sponsor a nationwide contest to see if locomotives were practical and, if they were, to pick the best one. The competition was held at Rainhill during a cold, drizzly week in October 1829. The challenge put to the contestants was to pull 20 tons at 10 miles per hour and to make 40 trips over a course a mile and a half long.

Ten entries were expected but only three appeared. Timothy Hackworth, the chief engineer of the Stockton & Darlington Railway, entered the *Sans Pareil*; George Stephenson entered the *Rocket*, soon to become the most famous locomotive in England; and Braithwaite and Ericsson entered their *Novelty*, which they built in less than 6 weeks to meet the deadline.

Weighing less than 2½ tons, the *Novelty* was smaller than the other entries. It was also the fastest of the three, reaching speeds of 28 miles per hour in the trials that took place the first day.[2] This was 4 miles per hour faster than its chief competitor, George Stephenson's *Rocket*, had managed during the opening session. On the second day the *Novelty*'s boiler became overheated and cracked. To reach it for repairs, Ericsson had to dismantle the engine. The steam-tight joints had to be resealed with a cementing compound that normally took a week to harden, but under the rules of the competition Braithwaite and Ericsson had to reenter the fray at once. To no one's surprise, when the *Novelty* reached a speed of a little over 15 miles per hour, the steam joints started to blow. This time the damage was major, and the *Novelty* had to retire permanently from competition. The *Sans Pareil* also failed to complete the course. Though the *Rocket* completed the trials at an unimpressive 15 miles per hour, it won the prize of 500 pounds sterling—and worldwide fame—by default.

Beyond the triumphs and failures of the machines and their engineers, the Rainhill contest achieved a spectacular and completely unexpected notoriety around the country. More than 10,000 specta-

[2]The *Novelty* would later become the first machine on wheels ever to move a mile in less than a minute (56 seconds, to be exact), but only after Stephenson had already walked off with the prize.

tors attended the meet, which convinced them and the British public of the practicality of steam locomotives. After Rainhill, newspaper readers in Europe and America began to follow railway developments with the same zeal that greeted the race to the moon 150 years later.

Railways across England were able to compete for the first time with canals for freight business. And railroads had so fired the public's imagination that railway companies were astonished to discover that the bulk of their revenues were soon coming from passenger traffic rather than freight. For the next two decades, England saw a railway boom, and the rest of Europe and America soon followed. More than any technology, steam engines harnessed to metal rails transformed Europe and America.

America and the Caloric Engine

The *Novelty* went on to greater successes and made Ericsson's a household name in Britain and across the Atlantic. This attention generated other engineering commissions. Ericsson became involved in the development of screw propellers for marine vessels, a more energy-efficient way of powering ships. In 1836 he patented a larger and more efficient propeller, first used in 1837. This device made it possible to transmit large amounts of power from a steam engine to propel the biggest oceangoing vessels, with huge potential impact on the world's navies. Based on this success, the U.S. Navy ordered a small iron vessel to be fitted by Ericsson with a steam engine and screw. It reached New York City in May 1839.

A few months later Ericsson immigrated to New York. He remained there, in a house on Beach Street in lower Manhattan (a stone's throw from the spot where Frank Shuman would later board the *Mauretania*), for the rest of his life. In 1848 he became an American citizen.

Ericsson continued tinkering with and exhibiting the revolutionary "caloric engine," the device he had come to England from Sweden to market and which he adapted in the United States to operate on solar power. The caloric engine was a hot-air engine. It operated on a simple principle. Ericsson had observed that air expanded when heated. He thought that by substituting air for steam in an engine, he might be able to cut back on fuel because air was already a gas, whereas steam had to be generated from water, a process that consumed energy. A hot-air engine was simply a device that took cool air into a cylinder, heated it, and used its expansive force to move a piston. As predicted by Ericsson (but in an era before the laws of thermodynamics were understood, to the astonishment of observers), the engine was much more efficient than a steam engine, running for hours on minuscule amounts of coal.

This efficiency was made possible by Ericsson's adaptation of what he called a "regenerator," his term for a heat exchanger. This device took heat from the hot exhaust air and transferred it to the cool incoming air. The regenerator was simply a cylinder filled with hundreds of wire disks made of a fine steel mesh. Steel is an excellent conductor of heat. The wire mesh changed temperature quickly and extracted most of the heat from the hot air as it left the engine, stored it momentarily in the metal, and then transferred it to the incoming stream of cool air.

It was this shift of heat in the exhaust to the air in the working cylinder that was responsible for the engine's high fuel efficiency. Waste heat from the machine was reduced, differentiating it from most steam engines, which threw off prodigious quantities of heat to the surroundings.

Any small external heat source—an alcohol lamp or a gas flame—put beneath the working cylinder could be used to start the caloric engine. Once running, the fuel simply made up the deficit of heat converted into mechanical energy or lost in the exhaust. When the external heat was shut down, the engine ran on its own for an hour or more just on the heat that was still trapped in the system. Though the hot-air engine was slow, its efficiency was much greater than any steam engine of its time, and its simple construction made it possible to use it in places where even a small steam plant—with boiler, engine, pumps, and condenser—was out of the question. Before the era of electricity, there was a large unmet need for a simple, foolproof fractional horsepower motor to run small devices like fans. The heat engine filled the bill.

Ericsson's invention was a commercial success in England and America. He sold 3000 caloric engines in less than 3 years, some with cylinders as small as 8 inches in diameter, others with cylinders as large as 32 inches across. Later versions of the caloric engine could be safely operated, it was advertised, by anyone who could light a lamp or a gas jet. Many powered small machines until electric motors replaced them in the years before the First World War.

In his early years in the United States, Ericsson adapted his caloric engine to run on solar heat. Though he sold a few, this adaptation was never the commercial success of its fuel-powered twins. The hot-air sun motor was useful at New York's latitude only in summer, and it could work only in direct sunlight. This setback prompted Ericsson to tackle the problem of energy storage. He liked the idea of compressing air in closed tanks for this purpose, but found that he had to use machinery and a reservoir so large that it dwarfed the caloric engine. He thought he might have better luck with electric batteries, but they, too, were too cumbersome to be useful.

Eventually he gave up and modified the caloric engine a last time so that it could be run using almost any combustible substance—coal, gas, alcohol, or wood—and he sold thousands of them. Delameter & Company, to which he eventually granted the patents, manufactured more than 50,000 of these pumps well into the twentieth century at prices ranging from $210 to $550, by far Ericsson's most successful invention from the point of view of his pocketbook. New York newspapers carried advertisements for them for years, extolling the "Ericsson Duplex Caloric Pumping Engine: No Boilers! No Steam! No Exhaust! Valveless Engine! Noiseless! No More Care than the Ordinary Cook Stove! Does Not Affect Insurance! Any Servant Girl Can Operate It! Absolutely Safe! Return Guaranteed If Unsatisfactory!"

Measuring the Intensity of Solar Heat

Ericsson spent his last 25 years in seclusion. He made no appearances at scientific conferences, though he kept up with everything that interested him in the technical press of England, France, and the United States. In the years after the Civil War, with hefty commissions flowing into his bank accounts, Ericsson felt prosperous enough to turn away from commercial engineering work.

He conducted experiments financed out of his own pocket—he would go through $100,000 of his savings before he was done—to determine the solar constant, and later to build solar-powered machines. From 1870 to 1872 Ericsson published nearly thirty articles in London's two newest science journals, *Engineering* and *Nature*, describing instruments he had invented for measuring the intensity of solar radiation.

With these devices (called solar calorimeters), Ericsson determined that the radiation reaching the surface of the earth was 5.64 Btu per minute per square foot. He estimated that about 79 percent of the solar energy arriving at the outer boundary of the earth's atmosphere made it through to sea level, giving him an estimate of 7.14 Btu per minute per square foot for the solar constant, which is close to the value settled on with more accurate equipment in the twentieth century.

Ericsson was notoriously secretive about the technical details of his solar inventions and wrote about them in generalities. In 1876 a great Centennial Exhibition was held in Philadelphia. Though his *Monitor* had made him world famous, Ericsson was not invited to participate. Miffed, he gathered his various publications together along with descriptions of his new inventions and compiled them in a great tome he called *Contributions to the Centennial Exhibition*. Three hundred copies of this work were printed by the Nation Press in New York at a cost to Ericsson of $30,000. The volume, which looks like an enormous, leather-bound family Bible, was a masterpiece of the printer's and the engraver's art. The illustrations (all drawn by Ericsson) were outstanding, and still fetch high prices when they turn up at auctions from cannibalized copies of the book.

Although he gave complete working details for his apparatus for measuring the solar constant in this record of his life's work, he did not provide technical specifications for his solar-powered machines, explaining that "experienced professional men will appreciate the motive, which is that of preventing enterprising persons from procuring patents for modifications." He had been burned by "enterprising persons" before, when earlier Ericsson patent applications, including several involving features on the *Monitor*, were copied and resubmitted, with slight changes, effectively preventing Ericsson from making and patenting (and profiting from) these small improvements himself. He therefore did not seek patent protection for his solar projects, though he did provide a general description of the apparatus he had designed, noting that it

> *consists of a series of polished parabolic troughs, in combination with a system of metallic tubes charged with water under pressure, exposed to the influence of converging solar rays, the augmented molecular action produced by the concentration being transferred to a central receiver, from which the accumulated energy is communicated to a single motor.*
>
> *Thus the mechanical power developed by concentrated solar heat is imparted to the solar steam engine without the intervention of a multitude of boilers, glass bells, gauges, feeders, etc. Moreover, the concentration apparatus, unlike the instrument of Mouchot, requires no parallactic motion, nor does its management call for any knowledge of the sun's declination from day to day. Its position is regulated by simply turning a handle until a certain index coincides with a certain bright line produced by the reflection of the sun's rays.*

Ericsson's boilers were efficient. He reported that "the mechanism which I have adopted for concentrating the sun's radiant heat abstracts, on an average, during nine hours a day, for all latitudes between the equator and 45 degrees, 3.5 Btu of heat per minute per square foot." Because he had estimated the amount of solar heat that reaches the earth's surface, under ideal conditions, to be 5.64 Btu per minute per square foot, the efficiency of Ericsson's boiler was more than 60 percent, which was remarkably high.

Ericsson's Solar Design

Ericsson had dismissed on theoretical grounds the use of Shuman-type hot boxes to gather up solar heat, pointing out that hot boxes simply could not concentrate solar energy enough to make the high temperatures and steam pressures needed to run conventional steam engines.

Ericsson machines had three main features: a conical mirror to concentrate the sunlight on a central axis (much like the stalk of an umbrella); a tube-shaped boiler in the focus of the mirror; and a small steam engine attached to convert the steam into mechanical power. Tests in 1870 showed that the boiler could provide enough power to run the small engine, but beyond this Ericsson said little more about this device. "Drawings and descriptions of the mechanism will not be presented," he wrote, "nor will the form of the generator be delineated or described." Later he added a tracking mechanism that kept the reflector aimed toward the sun as it moved through the sky. By 1875 Ericsson claimed that he had built and tested seven "sun motors."

Ericsson's secrecy provoked the public's curiosity. A Greenwich Village journalist, Charles Townsend Harris, had been following Ericsson's career since his Civil War naval exploits and made a practice of passing Ericsson's home in lower Manhattan on his way to and from work, hoping that someday he would be able to strike up an acquaintance with the "lonely, blue-garbed figure" he found seated on the stoop of the house basking in the morning sun or enjoying the evening breezes off the Hudson.

"Ericsson appeared to have no intimates, nor did his manner inspire intimacy," Harris wrote. He tried to gain the old man's sympathy by periodically giving him, without saying a word, envelopes of newspaper clippings about recent inventions—Ericsson was an avid follower of Thomas Edison's enterprises, among others. This eventually broke the ice. Harris was rewarded by an invitation to visit Ericsson's workshop, with a guided tour by his reclusive host.

Harris later recounted that he was led to the third floor of the engineer's house and into a room running the length and breadth of the building. Fitted out as a combination boiler, machine, and woodworking shop, it contained lathes, blocks and pulleys, an anvil, a full set of carpenter's tools, and other equipment. Harris noticed a strange-looking device standing in the corner. It appeared to be some sort of inverted, open umbrella, "but instead of cloth or silk, the divisions were occupied by long triangular mirrors which centered on a metal tube an inch or so in diameter and about 9 inches long."

This was Ericsson's first solar machine of 1870. Harris records that Ericsson "pointed it out with the pride of a parent introducing a first-born."

"Turned up as you see it, and exposed to the rays of the sun," Ericsson explained, "the light is concentrated on the metal tube containing water. The concentrated rays generate heat, acting like a burning glass. The water is converted into steam which, conveyed through the attached pipe, moves this little engine. In this size it may be regarded merely as a toy, its power depending on the small size of the generator, but the fact that it requires no fuel but sunlight, free to all, is its great advantage."

Ericsson's Largest Solar Machine

That solar machine was the first of Ericsson's prototypes (he would eventually build nine). A later model, demonstrated in 1872, used hot air instead of steam, patterned after his earlier caloric engine. Its pistons could push its flywheel at 400 revolutions per minute. By 1880 Ericsson had overcome his distrust of patents sufficiently to obtain one "for improvements in the solar engine."

This innovation—Ericsson's largest solar machine—took solar collection in a new direction that would have a tremendous impact on Frank Shuman's plans in Cairo 30 years later. It could concentrate sunlight eighteenfold and make steam at a pressure of 35 pounds per square inch and in enough quantity to operate a reciprocating steam engine with a 6-inch bore and an 8-inch stroke at 120 revolutions per minute.

To achieve this, he substituted parabolic troughs for the conical solar collectors. Unlike a true parabola (which focuses solar radiation

John Ericsson's solar motor, the first to use a "parabolic trough" collector (the curved shape that looks like a section of metal barrel). Frank Shuman later enlarged and adapted this design for his solar machine in Egypt.

onto a small area, or focal point), a parabolic trough is more like an oil drum cut in half lengthwise. It focuses solar rays in a line across the open side of the reflector. Ericsson believed this kind of reflector offered advantages over its dish-shaped counterpart: it was a simple, curved shape, less expensive to build, and, unlike a circular reflector, had only to track the sun in a single direction (up and down, if the device were lying horizontal, or east to west, if it were standing on end), eliminating the need for complex tracking machinery.

A drawback was that the device's temperatures and efficiencies were not as high as a dish-shaped reflector because the configuration spread radiation over a wider area (along a line rather than at a point), but this deficiency could be overcome by making the trough larger. Ericsson added a tube-shaped boiler—little more than a sturdy pipe painted black—and placed it in the focus of the trough. Positioned toward the sun and connected to a conventional steam engine, the machine ran successfully, though Ericsson refused to release technical data about the horsepower achieved.

In an article in *Nature* in January 1884, illustrated with his own engravings of his machines, Ericsson described his system:

The leading feature of the sun motor is that of concentrating the radiant heat by means of a rectangular trough having a curved bottom lined on the inside with polished plates [glass mirrors] so arranged that they reflect the sun's rays toward a cylindrical heater placed longitudinally above the trough. This heater, it is scarcely necessary to state, contains the acting medium, steam or air, employed to transfer the solar energy to the motor, the transfer being effected by means of cylinders provided with pistons and valves resembling those of motive engines of the ordinary type. Practical engineers as well as scientists, have demonstrated that solar energy cannot be rendered available for producing motive power, in consequence of the feebleness of solar radiation. The great cost of large reflectors and the difficulty of producing accurate curvature on a large scale, besides the great amount of labor called for in preventing the polished surface from becoming tarnished, are objections which have been supposed to render direct solar energy practically useless for producing mechanical power.

The device under consideration overcomes the stated objections by very simple means, as will be seen by the following description: The bottom of the rectangular trough consists of straight wooden staves, supported by iron ribs of parabolic curvature secured to the sides of the trough. On these staves the reflecting plates, consisting of flat window glass silvered on the under side, are fastened. It will be readily understood that the method thus adopted for concentrating the radiant heat does not call for a structure of great accuracy, provided the wooden staves are secured to the iron ribs in such a position that the silvered plates attached to the same reflect the solar rays toward the heater.

Referring to the illustration, it will be seen that the trough, 11 feet long and 16 feet broad, including a parallel opening in the bottom 12 inches wide, is sustained by a light truss attached to each end, the heater being supported by vertical plates secured to the truss. The heater is 6¼ inches in diameter, 11 feet long, exposing 130 × 9.8 = 1,274 superficial inches to the action of the reflected solar rays. The reflecting plates, each 3 inches wide and 26 inches long, intercept a beam of 130 × 180 = 23,400 square inches section. The trough is supported by a central pivot, round which it revolves. The change of inclination is effected by means of a horizontal axle, concealed by the trough, the entire mass being so accurately balanced that a pull of 5 pounds applied at the extremity enables a person to change the inclination or cause the whole to revolve. A single revolution of the motive engine develops more power than needed to turn the trough, and regulates its inclination so as to face the sun during a day's operation.

The motor shown by the illustration is a steam engine, the working cylinder being 6 inches in diameter, with 8-inch stroke. The piston rod, passing through the bottom of the cylinder, operates a force pump of 5 inches diameter. By means of an ordinary crosshead secured to the position and below the steam cylinder, and by ordinary connecting rods motion is imparted to a crank shaft and fly wheel, applied at the top of the engine frame, the object of this arrangement being that of showing the capability of the engine to work either pumps or mills. It should be noticed that the flexible steam pipe employed to convey the steam to the engine, as well as to the steam chamber attached to the upper end of the heater, have been excluded in the illustration. The average speed of the engine during the trials last summer was 120 turns per minute, the absolute pressure on the working piston being 35 pounds per square inch. The steam was worked expansively in the ratio of 1 to 3, with a nearly perfect vacuum kept up in the condenser enclosed in the pedestal which supports the engine frame.

In view of the foregoing, experts need not be told that the sun motor can be carried out on a sufficient scale to benefit very materially the sun-burnt regions of our planet.

The Impact of Ericsson's Work

The new Ericsson-style collection system became popular with later experimenters (and remains today one of the standards for modern plants—several of the largest solar-power plants built in Israel and California in the 1980s and 1990s opted for Ericsson-style parabolic trough reflectors because they continue to strike a good balance between efficiency, ease of operation, and low manufacturing cost).

Ericsson, with Adams, was also one of the earliest engineers to explore the political dimension of energy production and management:

> *Due consideration cannot fail to convince us that the rapid exhaustion of the European coal fields will soon cause great changes with reference to international relations, in favor of those countries which are in possession of continuous sun power. Upper Egypt, for instance, will, in the course of a few centuries, derive signal advantages and attain a high political position on account of her perpetual sunshine, and the consequent command of unlimited motive-power. The time will come when Europe must stop her mills for want of coal. Upper Egypt, then, with her never-ceasing sun-power, will invite the European manufacturer to remove his machinery and erect his mills on the firm ground along the sides of the alluvial plains of the Nile, where an amount of motive-power may be obtained many times more than that now employed by all the manufactories of Europe.*

In the last decade of his life, Ericsson refined his solar inventions, trying lighter materials for the reflectors and simplifying construction. He began to appreciate the possibilities of electricity for power generation. In December 1888, he designed his first high-speed (1300 revolutions per minute) steam engine, completed the next year. It supplied power for the next 12 years to an Edison generator in an electric lighting plant near his home.

By late 1888, he was so confident of the design and the practical performance of his most recent solar motors that he planned to mass-produce and sell the machines to the "owners of the sunburnt lands on the Pacific coast" for irrigation. By this time, he was immersed in solar experiments and had embarked on an ambitious program to build larger and more practical engines powered by sunlight, to make a "solar calorimeter" for measuring radiant heat as it emerged from the solar surface, and to invent ways to store solar energy overnight.

Ericsson's solar pyrometer, a device he perfected to measure the intensity of solar radiation in New York.

To be successful in America, he realized, the new devices would have to be cheaper than other sources of steam. Ericsson concluded that solar-powered devices, at least on the East Coast, would be more costly than conventional power because American-mined coal was so cheap in New York and Pennsylvania. In a letter to a fellow investor, he wrote: "The fact is that although the heat is obtained for nothing, so extensive, costly, and complex is the concentration apparatus that solar steam is many times more costly than steam produced by burning coal."

The extra expense for solar power was likely to be offset, he reasoned, only in remote tropical areas of the world and in the American West, locales that received large amounts of solar radiation and that also had to import coal from distant sources, driving up the price. It was an astute observation.

Unfortunately, Ericsson died the following year, in 1889. And because he destroyed most of his notes before he died, the detailed plans for his last sun motors died with him.

The search for a practical solar motor became something of a Holy Grail for American inventors after Ericsson's death, in part because of his own fame, in part because the editors of scientific press of the day had become convinced that solar power was likely to become one of the important new energy sources for the coming twentieth century. As the young century turned, there was to be an efflorescence of solar invention, especially in the United States. One of John Ericsson's greatest admirers, Frank Shuman, would be in its vanguard.

Though Ericsson's trough-style collectors were not part of Shuman's original design, they would soon be incorporated into the Ackermann and Shuman plan after Frank Shuman visited Egypt. Shuman was familiar with all of Ericsson's inventions, and Ericsson's status as one of America's most revered inventors in the post–Civil War era cannot have failed to have influenced the younger man from Philadelphia, reinforcing his interest in harnessing the sun's energy.

5

Solar-Powered Irrigation in Egypt

Frank Shuman sailed from England to Alexandria in the spring of 1912 to buy land, hire workers, and set the stage for the construction of a large solar power irrigation plant on the banks of the Nile, the first project to be underwritten by the Sun Power Company, the holding company he had previously incorporated in London. Shuman and his partner, A. S. E. Ackermann, had selected Egypt as the venue for their prototype because it was an important British possession where the high cost of imported fuel could give solar-generated power the economic edge it needed. It was also a nation that depended increasingly on irrigation for survival.

At the time of Frank Shuman's first visit to Cairo, British-occupied Egypt embraced some 350,000 square miles—an area more than twice the size of California—338,000, or 97 percent, of which was barren desert covered with a particularly fine-grained, gritty, choking red sand. It was a country apart from the rest of the African continent. The Egyptian desert isolated the populated regions of the nation as effectively as the Atlantic and Pacific Oceans isolated the United States. The nation's whole past and present were crowded into a narrow valley and fan-shaped river delta of which no more than 11,000 square miles could support permanent human habitation—an area less than one quarter the size of Frank Shuman's Pennsylvania.

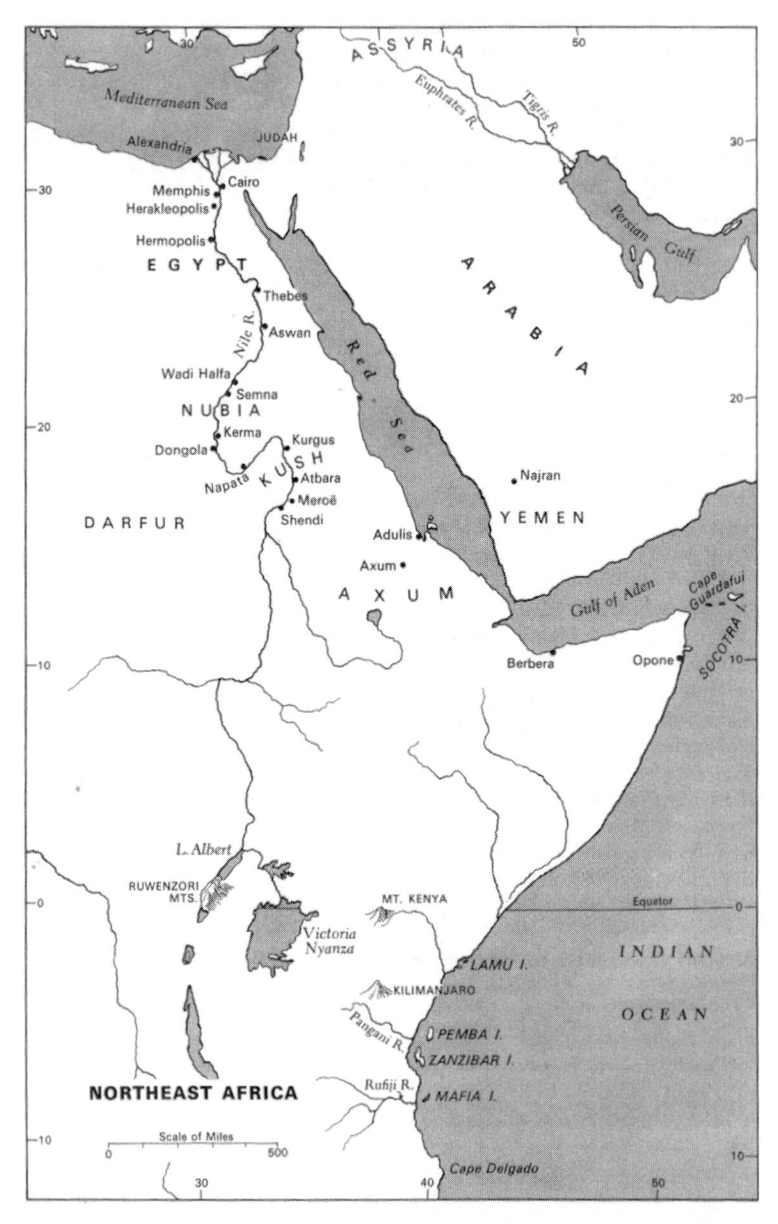

The River Nile, the "Great River" of the Bible and the world's longest, drains an area of over 1.1 million square miles, carrying fresh water from Lake Victoria (bottom) all the way to the Mediterranean Sea (top). The river's two main tributaries, the White Nile (from Lake Victoria) and the Blue Nile (flowing generally from Lake Tana in Ethiopia) come together to form the main body of the river near Khartoum before crossing the Egyptian desert to Alexandria.

This arena of Egyptian activity was coincident with the catchment basin of the world's longest river, the Nile. Since time began, the existence of Egypt had been dependent upon the great river—from Khartoum to Alexandria, the river dominated all things, and it was understood that men survived in Egypt only by grace of the Nile.

The heaviest rainfall in any part of Egypt was only a fifth of that which fell on the driest parts of the eastern portions of the United States, and that rain fell in the winter months alone and only along the Mediterranean coast. A hundred miles inland, at Cairo, the inhabitants were lucky to see four or five rain showers a year. In Upper Egypt, a rain shower was such a freak of nature that only the oldest living persons could usually recall one. The fact that rain had once fallen on Thebes was recorded by Herodotus as "a most remarkable prodigy."

Compounding the problem of low rainfall, consumption of water by agriculture was prodigious. Whereas a farm crop in Pennsylvania could thrive on the 4 to 10 inches of rainfall that fell during the whole of its growing season, in Egypt 10 inches of rainfall would merely supply the Egyptian cotton crop in August with the water it needed for 2 or 3 weeks. Agriculture in Egypt was not merely dependent on water, it required *abundant* water. On average, cultivated land soaked up 20 tons of water (or 20 cubic yards—they are about the same) per acre per day, and every drop of this water had to be carried from the Nile to the land by some artificial means.

The Source of Water in Egypt

As Frank Shuman would soon learn, the central fact of Egyptian life was that for all practical purposes every drop of water used in Egypt—whether by human being, by animals, or by plants—was carried down from the great lakes of central Africa and from the central Ethiopian massif by the Nile, including the water taken from wells. The politics and administration of Egypt centered on the politics and administration of the legendary river, and this was as true for Lord Kitchener as it had been for the pharaohs.

The Nile, 4200 miles long, flowed generally north from its most remote headstream, the Luvironza River in Burundi Territory, to enter the Mediterranean Sea through a vast triangular river delta in northern Egypt. The main body of the river was formed in Khartoum by the convergence of the Blue Nile, 1000 miles long, and the White Nile, 2300 miles in length. The Blue Nile arose from the summer rains that rushed into Lake Tana. It was the main source of Egypt's life-giving annual flood. The White Nile, rising in Lake Victoria in modern-day Uganda, had a more constant flow of water and provided the bulk of the great river's volume.

The river that flowed through Lower Egypt, passing Cairo to enter the delta, carried all but a small fraction of the water it had carried when it left Barbar, a village north of Khartoum beyond which no tributary added water to the main stream, some 1500 miles upriver. And yet by the time it reached the shores of the sea, very little water from the Nile passed into the Mediterranean, at least not in the summer months. It had all been diverted for irrigation, not a drop wasted. Through this long journey, after the Atbara River joined it 30 miles above Barbar, the Nile did not receive so much as a trickle of water from any source. Instead, it was continually losing water into the soil and into the air.

Once every year after the rains in Ethiopia, the Nile rose in flood. The high-water mark at any point along the river varied enormously, depending on the local configuration of the irrigation canals humans had devised over the centuries. The natural flood at Maadi, the little community south of Cairo where Frank Shuman would set up his solar engineering works, usually began when the surface of the river had fallen about 17 feet below the level of the surrounding plains. During four or five weeks starting in June, the water level would rise just to the river's banks. The swirl and rush of the chocolate-brown water—110,000 cubic feet of it per second—passed under the Cairo bridges at the speed of a fast long-distance runner. To see this great sheet of water racing across one of the driest deserts in the world was to understand the terror and reverence the Nile had inspired from the beginning of Egyptian history.

But the annual flood was only half the magic of the Nile. If no water were to follow the rains of Ethiopia, the great river would dry up before it reached Cairo in the summer months, and the whole of Egypt would die of drought. The continued existence of Egypt was dependent upon the double origin of the Nile, which was fed from two great catchment areas—the Great Lakes of central Africa and the plateau of Ethiopia. Roughly, it was the former that kept the Nile flowing all year around while the latter alone lifted it into flood.

Contrary to the popular belief that prevailed in the nineteenth century, it was not the Great Lakes of Uganda that provided the river with its flood waters, although there was a slight summer increase in volume every year from that source. The cool mountain plateaus of Ethiopia that intercepted the humid summer winds blowing west over the Indian Ocean in summer, becoming drenched in torrential rains, captured this additional gift of water. There were no large basins to steady the runoff from the Ethiopian mountains, although Lake Tana, usually regarded as the source of the Blue Nile, provided some tempering of the flow.

A dry creek in winter, the Blue Nile became a raging mountain torrent in summer, its level rising several feet a day when the rains began. As the river left Ethiopia for the dusty plains of Sudan at Roseires, the average change in the depth of water between low-stage and flood-stage was 30 feet. The lakes of Uganda might provide, through the White Nile, the steady flow that made up the better part of the water in the great river year around, but it was Ethiopia's Blue Nile that provided the life-giving annual deluge. By the time the combined rivers left Khartoum, the waters had steadied to a uniform flow that would be more or less uninterrupted for the 1500-mile remaining journey across the African continent to Cairo.

Choosing the Site

Maadi, the site Shuman chose for his solar power works, occupied a slightly elevated area on the banks of the Nile south of the capital, just before the point where the river entered the city. Though recent excavation had shown that Maadi was settled in ancient times, the modern community dated from 1904 when a railway track was built connecting Cairo to Helwan, a village 12 miles south of the city. The attraction of Maadi for Shuman was twofold: First, it was a farming community near Cairo and the administrative center of the country and therefore close enough for senior British officials to visit easily;

by 1911–12 British colonial administrators who wanted to escape the urban congestion of Cairo were building pleasant villas in Maadi. And second, it was just high enough above the river to make gravity-fed irrigation difficult.

The British Agency garden overhanging the river was among the first sights Frank Shuman would have seen as he passed Maadi on his first riverboat journey into Cairo from the southern desert. The agency was the center of power in the country, and the office and home of Lord Kitchener. Cairo itself loomed above the Nile's eastern bank on a series of low hills. Beyond it, the river forked left and right to enclose the Gezira quarter with its cool British villas and the Sporting Club, the only field-size acreage in the whole of Egypt that boasted a well-tended grass lawn.

Past the northern port of Cairo, the river flowed into the delta itself as the low hills of the pyramid plateau to the west dropped out of sight. Here the land was only 60 feet above the level of the sea at Alexandria, 100 miles to the north. In this triangular plain of the delta, there was hardly a molehill to break the monotony of flatness. It was an ocean either of floodwaters or of crops, depending upon the season. Apart from providing for the small needs of the villages along the river in Upper Egypt, the whole of the Nile's bounty was used here to provide irrigation. In this triangle with 100-mile sides, four millennia of Egyptian ingenuity had conspired to prevent the river's precious water from escaping into the sea.

The Nile Delta was densely settled. In 1910 the delta had a population of 6 million (apart from Cairo, the largest city in Africa at the time, with its own urban population of 700,000) on a cultivated area of a little more than 3 million acres. This land produced annually about 20 million pounds sterling worth of the world's finest cotton, besides enough food to support the Egyptian population itself. As an agricultural country devoid of industry, Egypt depended wholly upon the export of cotton, though it was completely self-sufficient in production of food.

It was on the southern periphery of this great, fertile river delta that Frank Shuman leased acreage not far from the banks of the Nile to build his solar power demonstration project. The area was surrounded by richly fertile land only a portion of which the local fellahin irrigated with backbreaking human and animal labor. In the first weeks he spent in Egypt, Shuman surveyed these local irrigation methods carefully to determine with precision how much water each could raise from the Nile to the fields and the extent to which sun-powered pumps might extend the arable acreage near the waterworks he proposed to build.

Irrigation along the Nile Valley

Apart from imported inventions like the Archimedean screw and coal-fired water pumps, the Egyptian farmer had only four basic methods for lifting water along the Nile valley, even in places where free-flow irrigation was possible part of the year. These methods were all labor-intensive, exacting a heavy toll on Egyptian workers. According to the work they had to do and the frequency with which waterlifts needed to be manned, the Egyptian farmers relied on the *nattala*, the *badala*, the *shadoof*, and the *sakieh*, all ingenious devices whose first recorded use is documented in hieroglyphs from the dawn of history.

The *nattala* was simply a basket swung by two men with ropes with which they scooped up the water in which they stood up to their knees, lifting it over a low bank in front of them. The *badala* was similarly used for low lifts, consisting simply of a tipping trough worked by one man, pivoted at its open end.

By 1911 Shuman observed that both these implements, though still in use, had largely been supplanted by the Archimedean screw. This was a wooden cylinder hooped with iron staves about 10 feet long and 18 inches in diameter containing a spiral partition. The cylinder was mounted to revolve on its axis, its lower end dipping into the water to be raised. When the cylinder was rotated by hand, the water was carried up into the cylinder by the screw and poured out the other end. It was not a cheap implement, costing the fellahin about as much as the local plows, and it was a feeble pump, lifting a thin stream of water about 2 feet. One man could operate it all day. It was more efficient than either the *nattala* or *badala*, and much less exhausting to work, and therefore favored by those who could afford it.

A modern-day *shadoof* lifting water on the Nile.

For higher lifts the simplest appliance was the *shadoof,* a long counterpoised wooden boom, 8 to 10 feet above the ground, pivoted between two pillars with a bucket at its long end. The fellah operator pulled on a rope that lowered the long end of the boom so that the bucket submerged and filled with water. The fellah then released the rope, allowing the counterweight to raise the bucket to the desired level, then dumped the water in the higher channel. Then the cycle repeated. If a young boy was available in the family, he might be conscripted to run to and fro along a plank nailed to the upper side of the boom to provide a living, movable counterpoise to help his father or brother power the lift, running toward the pivot when the bucket was empty and then away toward the far end of the boom when the bucket was full, helping to raise the water with the weight of his body.

The leather dipper of a *shadoof* held about 4 gallons of water. It was an ancient and remarkably efficient machine whose only disadvantage was that it required human muscle to power it. The *shadoof* worked best on a lift of about 6 feet. Higher lifts could be worked by using two or three *shadoof,* each with its own human operator,

A camel-operated *sakieh* on the banks of the Nile in Egypt. This British engraving was made about a decade before Shuman's experiments in Maadi. Note the clay jars secured to the endless loop of rope used to lift the water from the river's surface.

arranged in sequence along a tier—a kind of reverse cascade. The maximum discharge of a *shadoof*, Shuman calculated, was about 8 tons of water per hour (2000 gallons), which required about seven strokes of the boom per minute. To maintain this rate, two or more human operators were needed to work in tandem, for even the brawniest man could not maintain such backbreaking exertion very long without a rest.[1]

The creaking groan of the great wooden cogwheel of the *sakieh* was another familiar feature Frank Shuman found in the fields at Maadi, a sound that had been heard in Egypt since pharaonic times.

[1]The average man can deliver muscle power equivalent to one-seventh of a horsepower, or about the energy needed to keep a 100-watt bulb brightly lit. For short periods of great exertion, some men can deliver about twice that much.

The *sakieh* was a waterwheel with buckets attached to its periphery or to an endless loop of rope, usually arranged over a pit sunk in the fields down to the level of the water-bearing gravel. A draft animal turned the wheel—a camel, an ox, or a buffalo.

The amount of water that could be raised with a *sakieh* was substantial—easily 30 tons of water (7,200 gallons) an hour or more—but highly variable, depending on the speed of rotation and the number of buckets in the loop. A *sakieh* was a major investment, used for lifts of up to 25 feet. Usually groups of farmers would pool funds to meet its cost, both in terms of the animals that had to be fed and cared for to provide power and for the equipment.

A variation on the *sakieh* that Shuman observed was the *taboot,* which replaced the drum and buckets with a wheel with boxes in its rim. Another was the *noria,* which was a modern form of the *sakieh* made of iron. The latter, it turned out, was less reliable than the wooden form because its teeth were liable to strip when the drowsy draft-animal leapt forward in response to a blow from the stick of the small boy charged with making it trudge in endless circles.

Investors Ask for a Second Opinion

By the beginning of the spring of 1912, Frank Shuman had made the preliminary arrangements necessary in Egypt to set up a demonstration solar-powered steam plant at Maadi. He had rented the land, hired local workers, and made arrangements with local shipping and transit agents to receive cargo from Britain and the United States—the crates of equipment he planned to prefabricate at home and assemble on site.

He had been staying at the elegant Shepheards Hotel in the posh Garden City quarter of Cairo. It was the grandest hotel in the city. In its palatial dining room were assembled daily some 200 or 300 travelers of all nationalities—mainly British and American, with a sprinkling of German, Belgian, and French—about half of whom were Anglo-Indians homeward or outward bound or European residents who were spending three months in Cairo to avoid the cold of a European winter. The other half were businessmen seeking connections of various kinds in the capital, or travelers intent on making the voyage up the Nile for health or recreation.

Shuman was able to make most of the European contacts he needed in Egypt without ever leaving the hotel—civil servants, bankers, journalists, investors, landowners, anyone at all who might help his solar venture succeed. Before he had been there 2 weeks, he knew everybody's name and business, from the invalids in search of an Egyptian cure, artists in search of Egyptian subjects, sports enthusiasts eager to bag a crocodile, statesmen out for a holiday, press correspondents alert for gossip about Egypt for the London papers, collectors of antiquities, and other engineers and businesspeople like himself who were attracted to Cairo by its prewar growth and bustle and the many opportunities present there for making quick money.

As he was preparing to abandon the hotel for leased housing in the bucolic residential suburb now growing up around Maadi, he received a letter from A. S. E. Ackermann in London, addressed to Cairo, much traveled and with eight transit date stamps, a letter that had apparently nearly been lost several times in its journey. Ackermann said he had tried to telegraph without success. The news was not good; there had been a few snags in London.

It seemed that the Egyptian solar project had become a favorite topic of conversation in the elite clubs of London. Every armchair engineer was eager to weigh in with his own opinion on its merits—and to give advice on how to make it better. Swayed by this talk, some of the more prominent British investors now wanted a second opinion on the solar absorber. They had demanded a review of the project by a prominent British scientist, Professor C. V. Boys. Could Shuman interrupt his important work in Egypt, Ackermann asked, to return at once to London? Only Shuman himself could sort out the problems caused by this new development, Ackermann believed.

The name of Charles Vernon Boys was certainly one known to Frank Shuman. Boys was one of England's most eminent physicists and designers of sensitive scientific instruments, known especially for his use of the torsion of quartz fibers in the measurement of minuscule forces of nature. He had applied this technique in his "radiomicrometer" for measuring radiant heat and also in his elaboration of Henry Cavendish's famous experiment to measure Newton's Constant of Gravitation.

Boys's early fame (he was born in 1855, 5 years before Shuman) as a young assistant professor of physics at the Royal College of Science, South Kensington, rested mainly on his efforts to measure the constant of gravitation first described by Isaac Newton, a force usually denoted in equations by the capital letter G. He accomplished this by using a phenomenally accurate balance that used a tiny fiber and a torsion beam only a fraction of an inch long to measure the gravitational attraction between a pair of solid gold balls. He was able to measure this force—very, very weak though it was—to about 1 part in 1000, which was regarded as a great achievement by his peers. In 1888, at the tender age of 33, Boys was elected a Fellow of the Royal Society, the highest honor to which an English scientist could aspire (besides knighthood—that accolade would follow in 1935).

Boys's scientific opinions carried great weight in Great Britain, and Frank Shuman could ill afford to ignore any views he might have on the solar demonstration project, especially as Sun Power Company investors had requested it. Sitting on the great veranda of Shepheards Hotel watching the sun set on the Giza plateau, Frank Shuman quickly determined that if Boys wanted to comment on the soundness of the proposed solar venture, he would have to be given a respectful hearing. And Shuman would certainly have to be present himself to receive it in London. There was nothing else to be done. With some irritation at having been interrupted just as his efforts appeared to be bearing fruit in Egypt, Shuman cabled Ackermann that he would be on board the next steamer out of Port Said.

Kitchener's Irrigation Projects

In the months that Frank Shuman was studying the irrigation systems of Egypt and establishing the groundwork for his solar steam plant, Lord Kitchener of Khartoum was rediscovering Egypt. Unlike

Shuman, Kitchener was no stranger to the lands of the Nile. In 1892 he had been appointed commander-in-chief of the Egyptian Army, completing its reorganization in time for it to take part, in 1898, in the victory that would win him permanent fame, the Battle of Omdurman. That engagement near Khartoum restored British prestige in Egypt and Sudan and reasserted British jurisdiction over Sudan.

Kitchener had determined early in his new Egyptian tour that he would stay out of politics as much as possible and focus on practical projects whose success was more tangible. He decided not to concern himself publicly with constitutional affairs, preferring to focus on agriculture, the building of roads, industrial development, reform of the legal system, and improved irrigation. Irrigating, draining, and reclaiming land was very much on Kitchener's mind, and he went about this work with the vigor and single-mindedness he could not safely have brought to Egyptian politics. For a man of such a strong practical bent and endowed with supreme self-confidence, Egypt provided a theater in which to exercise power in practical matters on a grand scale.

In Egypt, Kitchener proclaimed, "prosperity and water go hand in hand." Kitchener was convinced that "the proper maintenance of those great Public Works [the water infrastructure] is the first condition of Egypt. . . . Without an elaborate canal system it might become again a desert."

From its southern confines near Wadi Halfa and Aswan all the way down to Cairo, the agricultural land of Egypt was still a narrow but now gradually widening belt of green. The most fertile fields were mainly on the western bank of the Nile. Nowhere were they more than 10 to 15 miles wide. Under the old, gravity-fed system of Egyptian irrigation, the country was divided into large basins. Into these basins the Nile, in flood, was allowed to flow with its fertilizing, silt-laden, life-giving waters. This natural system of irrigation meant having crops in the soil from August to March. It did not allow production, in most parts of the Nile Valley, of the more valuable cotton, which in Egypt grew between March and October. It was this that Kitchener was determined to change.

To serve the cotton crop upriver from the delta, perennial irrigation had been introduced to raise water out of the river using all the backbreaking implements Shuman had surveyed. In so-called perennial irrigation, the Nile was not allowed merely to flood the land and lie stagnant for a month or more. Instead, Egypt was divided up by canals to provide each portion of ground with a permanent supply of water, permitting farmers to take as much water as needed for each field.

British planners had tackled the problem of increasing the Egyptian water supply. The first step taken was the simple but very effective one of holding back a part of the Nile flood and storing it for use the following summer at low water. This was done by building a dam across the Nile Valley at the First Cataract near the village of Aswan. The choice of the site was dictated by the river having, at that point, cut its way through a bed of granite that could serve to anchor the dam. Below Aswan, finding bedrock beneath the river was nearly impossible. Engineers dug for hundreds of feet without finding anything but gravel and rotten bedrock that could not support pylons.

At Aswan, the entire river—sides and the bottom—was contained by a single, colossal mass of solid granite. The original dam, raised in 1902, was not as high as that originally proposed by its architect, Sir William Willcocks, because of an outcry about potential damage to the antiquities on the island of Philae and other historic sites upstream. Later, when the threatened areas had been carefully surveyed by archaeologists, the dam was raised to a height of 143 feet above the valley floor, holding up the level of the river behind it nearly to the Sudan border and storing some 2.5 billion tons of water (600 billion gallons) for irrigation. The dam building and lake expansion had been proceeding on schedule in the decade before Kitchener took command of Egypt's administration.

The year before his appointment, Kitchener took a long vacation by steamer upriver from Alexandria. He was given a huge reception in Khartoum, the city he had delivered from Mahdism. There, he inspected every irrigation project he could find. He borrowed another steamer from his friend and protégé the governor of Sudan,

Sir Reginald Wingate, to take his party all the way up the White Nile to Entebbe on the shores of Lake Victoria in the heart of Africa.

This trip and many others he took along the river after his appointment made Kitchener familiar with all the engineering work being done along the Nile. His challenge was to find a use for all this water by increasing arable land to grow crops downriver. To do that required moving water, vast quantities of it, to lands previously too high to benefit from Nile floods. Kitchener saw that one way to achieve this was to place pumps at the head of the main canals and lift water into them. This strategy was advocated by many in Egypt. But it was expensive and suffered also from the liability of being dependent on supplies of imported fuel. Because Egypt was not a coal-producing nation, a pump-fed system would be vulnerable—it could be brought to a standstill if the supply of imported coal was somehow disrupted, say, in time of war. But, in those rare instances where pumps had been installed, no one could argue with the results this costly method produced, for they were simply spectacular. That was the story of Qom Ombo, and it was a lesson Kitchener learned well.

To see parts of the country away from the capital, Kitchener frequently took semiroyal tours of the provinces in a special train to inspect irrigation facilities and other improvements he had ordered. He spent hours gazing from his special railway coach on the green wealth of the Nile, the increase of which he hoped would be his legacy to the country. On one of his earliest in-country tours, Kitchener visited the Qom Ombo plantation in Upper Egypt, about 40 miles below the dam at Aswan. With the completion of the Aswan Reservoir and its subsequent heightening in 1909, several hundred thousand desert acres were brought under cultivation for the first time. The extra water during the dry season allowed for perennial irrigation so that a second crop could be planted each year. This led to a doubling and then a tripling of land values around the dam, the allocation of the new acreage to sugar cane, and the burgeoning of sugar factories all over Upper Egypt.

Sugar production would eventually become the single largest industrial employer and sugarcane the nation's second largest cash crop (after cotton). But these benefits accrued only to those lands below the water level of the new reservoir. Unfortunately, vast areas of otherwise fertile land were above the reach of gravity-fed irrigation canals or too far from Aswan to benefit directly. Qom Ombo, so far north of Aswan, was beyond reach of the life-giving waters of the reservoir.

Qom Ombo was a wide, flat valley that ran eastward into the desert, about 20 miles to a side—400 square miles of fertile land that lacked only water to become "an agricultural Eden" in the eyes of British administrators. When Kitchener had last served in Egypt, as head of the Egyptian army, Qom Ombo had been a howling desert bare of any vegetation and with a daily temperature in the shade of about 115 degrees Fahrenheit. Travelers by rail (the railroad connecting Luxor with Aswan ran through it) regarded it as one of the most unpleasant stretches of the journey to Aswan, the desert glimmering with mirages and torn by howling whirlwinds.

Then a British group formed a company to bring it under cultivation. To irrigate the initial tract of 20,000 acres to grow cane sugar, every drop of irrigation water had to be lifted between 35 and 65 feet above the river. Three huge pumps of 1500 horsepower apiece were set up along the banks of the Nile, each one capable of discharging a ton of water (240 gallons) per second into the canals above. This was the equivalent in water movement per pump of between 2500 and 5000 human workers manning *shadoof*, depending on the height the water was raised, or from 170 to 340 buffalo-powered *sakieh*.

The diameter of the main feed pipe carrying water from the river was so large that horses could be led through it. The Qom Ombo embankment was so steep that the main canal irrigating the fields had to be brought toward the pumping station on a berm, its bed reinforced with steel plates for a half mile until the level ground fringing the edge of the plain was reached. An observer described the construction as "suggestive of giants playing at mud-pies."

The results of this experiment in machine-powered artificial irrigation of the Egyptian desert were spectacular. The barren land was transformed into an expanse of green fields centered around a new town with its own sugar factory, fed now entirely by its own agricultural production. This was progress, but it came at a cost. The steam engines powering the giant pumps were fueled by coal—coal carried by steamship 3500 miles from Wales and then by barge or rail 700 additional miles to Qom Ombo. This precious coal cost the sugar cane plantation owners the equivalent of over $40 a ton at the furnace mouth.

Surely there was a better way, Kitchener and his advisers mused, to power these life-giving pumps, making mechanical, pump-assisted irrigation more economical up and down the great river.

Redesigning the Heat Absorbers

Back in London, Shuman quickly divined the new direction in which his financial backers wanted to take the construction of the Maadi plant. It would prove to be a change he was powerless to resist, and its effects would be immeasurable. With the U.K. financing had come stipulations. Shuman was obligated to let his backers bring in Charles Vernon Boys to review the design of the demonstration plant in Egypt and suggest possible improvements.

The famous physicist had already recommended a series of radical changes, first among them a fundamental shift in the design of the solar-heat absorbers. Boys's concern seemed to be that the simple asphalt design was not up to the job. He wanted to increase the light-gathering capacity of the absorber by substituting parabolic troughs (much like Ericsson's) for the field of asphalt. To Shuman's annoyance, Ackermann seemed swayed by Boys's analysis, adding his own technical suggestion that if the parabolic troughs were to be truly effective, they would have to track the sun across the sky throughout the day.

Shuman listened stone-faced to all the arguments. For him, it was a question of solar-heat concentration and its tradeoff with cost. There was never any doubt that techniques existed to produce higher concentrations of heat than an asphalt field would provide, but were they worth the extra expense? Shuman now feared that the delicate balance he had established between performance and cost—keeping costs down by favoring simple technologies—would be upset, with the Maadi plant careening off toward greater complexity and expense. His central commitment, that the system be uncomplicated and inexpensive, was at risk of being compromised or abandoned.

Boys had done his homework. He had reviewed all of Shuman's work from 1906 on, beginning with the initial 3½-horsepower machine fed by 1200 square feet of flat, fixed hot-box collection area covered with a single layer of glass. In that model, rows of parallel horizontal black pipes containing ether were coiled in the hot box and heated by the surrounding water. That experiment had demonstrated the theoretical possibility of converting sunlight into mechanical power on a larger scale while revealing the limitations of ether machines.

In subsequent models, Shuman had added (reluctantly, it seemed) mirrors on either side of the hot box arranged in such a way that 6 square feet of solar rays were concentrated on 3 square feet of hot box—creating a concentration ratio of 2 to 1. Boys was captivated by this design, and his suggested modifications were largely an extension of this principle. He was concerned that only the surface of the hot boxes collected solar energy, noting correctly that the bottom of the absorbers, made of asphalt, lost some heat to the ground below.

It seemed to Boys that rather than go back to a simple hot box with no mirrors, the proper direction was to elaborate the use of mirrors in a way that would increase solar concentration. He suggested that the mirrors surround a central, tubular boiler so that the sun heated every square inch of surface area on the boiler. Though his British colleagues did not at first see the parallel, this was essentially the same parabolic trough design that had earlier been advocated by John Ericsson in the United States.

It was apparent to all three men that while a parabolic trough collector would increase the amount of heat provided to the boiler, its cost would greatly exceed that of the simple asphalt, wood, and glass design favored by Shuman. Shuman felt compelled to make the case once again for the flat hot boxes without mirrors or other costly concentration devices, noting that a similar design had been able to heat water sufficiently to run a small engine even when snow was on the ground in Tacony. It would certainly be effective, he declared, in the oven that was the Egyptian desert. Boys seemed not to be swayed.

Having made his arguments as best he could over a period of several days, Shuman realized that he was losing the debate. His British backers were adamant that Boys be brought on board as a consultant to the project and that his ideas be reflected in a new, improved design. Ackermann, who initially thought the Boys design was original, tried to put a diplomatic face on the new contentiousness within the Sun Power Company, papering it over while, perhaps as a sop to Shuman, gently taking Boys to task for not having credited Ericsson properly for the parabolic trough concept.

The main impact of the proposed change was that the cost of the solar collection area would now rise significantly from the $0.25 per square foot budgeted for the asphalt model. The $40,000 that formerly would have bought 160,000 square feet of absorber, sufficient to power a 1000-horsepower steam turbine, would now pay for considerably less heat absorption. Although each square foot of reflector would presumably be more efficient, the cost differential was such that Shuman realized he would have to scale down the size of the steam engine. There was no way $40,000 of parabolic reflectors could support a 1000-horsepower machine, no matter how efficient they were.

Barring another round of fundraising (which Shuman was loathe to undertake without initial results to discuss with the first round of supporters), the Sun Power Company would be obligated to work within the budget that had already been drawn up for Egypt. Given the increased cost per unit area of the absorber, the demonstration project would therefore have to be scaled down significantly. The character of the steam engine would also have to change. Because small turbines were not as efficient as reciprocating steam engines, the scaled-down plant would require the latter, precluding the possibility of using a turbine, the engine design Shuman preferred.

Although the consequences of these changes on the future Maadi plant could not be fully foreseen, Shuman believed it continued to be important to keep within the broad economic guidelines he had originally laid out.

"Sun power plants, once constructed, dispense with all fuel," he wrote, while in order to be practical and commercially profitable, they must conform to the following requirements:

> *They must not cost so much to construct that the interest on the cost over and above that of a coal-burning plant of equal capacity will annul the profit made by the saving of the cost of the fuel.*
>
> *They must be constructed of such material and in such a manner that few repairs are needed, and so that they will last very many years.*
>
> *They must be constructed strong enough to stand the heaviest gales that may occur in the localities where they are erected.*
>
> *They must be sufficiently simple for anyone running an ordinary coal-burning plant to be able to run them.*

All four of these axioms were potentially challenged by the new direction the design of the plant was taking. Though mass production of the reflector troughs in the United States or Europe might cut down on their unit cost, the reflectors were clearly more complex than the original design, more fragile, and potentially more difficult for workers with little training to assemble on site and to operate.

Boys believed that the entire sun-collecting framework could be constructed out of prefabricated structural pieces manufactured quickly and cheaply in any competent British machine shop. The mirrors lining the troughs were to be made of ordinary plate glass, silvered in such a way as to keep their cost down to 5 cents per square foot. The foundations were to be made of poured concrete. The engine, pump, and auxiliaries were the same, except for scale and the switch from turbine to reciprocating engine, as in the original design. These components were comparable to those found in any conventional coal-burning plant, except that cylinders in the steam engine would be larger than normal because it had to operate at low pressure and needed the extra surface area to achieve efficiency. The parabolic heat absorbers were to be constructed entirely of concrete, steel, and glass—all materials that had been tested and used time and again in the tropics.

Shuman anticipated that the exposed metal would have to be painted every 5 or 6 years to control rust but would otherwise not require any special treatment. The troughs would be tested to withstand safely a wind pressure of 30 pounds per square foot, deemed to be above the upper limit of storms in Egypt.

Even Shuman had to admit that these new parameters were no bar to substituting the modified Boys design, though they were more complex and more costly than the original specifications. What had swayed the British investors was the superior ability of the parabolic troughs to concentrate sunlight—achieving a ratio of 5 to 1. The new conception of the plant involved five sun-absorbing troughs, each 13 feet 4 inches wide at the opening presented to the sun and about 205 feet long. The parabolic trough absorbers would present 13,666 square feet of surface to the Egyptian skies to capture the sun's rays, as opposed to the 160,000 square feet of the proposed asphalt absorber, though with a 5-to-1 concentration advantage.

Total energy collected, however, would be substantially reduced. The parabolic trough could capture only 8.5 percent of the energy of the larger design, even though it would concentrate that energy much more effectively and achieve more rapid heating of the water, and higher water temperatures. But the overall size of the experiment had now to be considerably diminished. Instead of a 1000-horsepower machine, the new arrangement would likely barely support 85 horsepower.

It was a huge departure from Shuman's original conception, and he may have rued the day he decided to seek British financing for his Egyptian project—considering all the strings attached to such support. What was it that had so seized the imagination of C. V. Boys and the British financiers who supported him to take Shuman's Egyptian project in this new direction? Why abandon the elegant simplicity and the grand scale of the 1000-horsepower machine for a device that was much more complicated, more expensive to build, smaller, and more likely to break down?

In the end, it is likely that C. V. Boys's interest in solar concentrators could be traced back to the very first sun machine, a device that had influenced all later solar inventors, the machine developed nearly 60 years earlier in France by Augustin Mouchot. That was a machine that had also been tested in the North African desert, though in Algeria rather than Egypt, and was as influential in its own way as Shuman's would be.

Boys believed that it was always wise to build on past successes, although he may not have realized that in emulating the attractive features of Augustin Mouchot's concentration concept of a solar engine, the Maadi plant would now be handicapped by its less-advertised weaknesses. He had likely not studied the failures of Mouchot as thoroughly as he had studied his successes.

6

Augustin Mouchot and the First "Sun Engine"

The work of Adams, Ericsson, and Shuman had been directly influenced by the solar conceptions of Augustin Mouchot, a man who arrived on the scene in nineteenth century France at precisely that moment when his ideas were likely to attract the most attention. It was a time when French industrial might was at a peak and her leaders open to new ideas, none more so than her emperor. In 1867, to commemorate the explosion of technology that had accompanied the industrial and artistic carnival over which he had presided for 15 years, France's Napoleon III decided to invite the whole world to an international exposition that he would host in Paris.

The focus of the exhibition, a few yards from where the Eiffel Tower stands today, was a vast elliptical building of glass, 1500 feet long set in a filigree of ironwork, not unlike London's own Crystal Palace. This Parisian castle, a gigantic oval that covered more than 35 acres on the bleak marching fields of the Champs-de-Mars, appeared from the outside like a gigantic steel tank with a 1-mile circumference set with hundreds of thousands of panes of glass. An open garden graced the middle, and 12 concentric aisles separated the displays. Walking around the massive exhibition was supposed to give visitors the impression of circling the globe.

So high was the dome, marveled a visitor, "that one had to use a machine to reach it, and the roof with its red arcades breached by the blue of the sky gave you a sensation of the immensity of the

Coliseum." Inside this pavilion, all the leading countries of the new industrial era flaunted their exhibits, showing the heights attained by their civilizations.

The pavilion was divided into seven sectors, each representing a branch of human enterprise. Among their exhibits, the Americans, recovering from a brutal civil war, sent a complete field service or *ambulance*, as it was called, representing the epitome of military medicine. The throngs passed it by, showering their attention on a patented new piece of American furniture described as a *rocking chair*. Britain sent locomotives and giant stationary steam engines. There were displays of a new featherweight "wonder metal" called *aluminum*, so precious that the emperor himself had ordered a special dinner service made of it for the Empress Eugenie. In the science section, which comprised the nucleus of the exhibition, there were marvelous products made from the newly discovered substance known as *petroleum*. From Prussia had come an immense 50-ton gun

One of the largest Mouchot devices ever built on display at the Universal Exposition in Paris in 1878 on the banks of the Seine. It was a prototype of this device that so intrigued Napoleon III in 1867 and spurred him to provide Mouchot with financing.

exhibited by Alfred Krupp, the "cannon king" of Essen, who had launched his career as a manufacturer of railway wheels. Firing a 1000-pound shell, it was the biggest gun the world had ever seen.

As the weeks went by, the famous poured into the City of Light from every corner of the globe. Even jaded Parisians gawked at the resplendent monarchs and their courts, from England, Egypt, Turkey, Greece, Sweden, Denmark, Belgium, Spain, Japan, Prussia, and Russia. Rarely had so much royalty congregated in one place. Walking through the exhibition reminded Parisians of how small the world had become as a result of the telegraph, the steamship, railroads, and the soon-to-be-opened Suez Canal.

In the midst of this efflorescence of industry and art loomed a fantastic mechanical object that towered above the other exhibits. To modern eyes, in an engraving from the period, it appears improbable—a huge metal ray gun pointing skyward, or perhaps a big satellite dish perched along the banks of the Seine a century before its time. To contemporary Parisians it was "a mammoth lamp-shade, with its concavity directed to heaven," in the words of one journalist, "a steel sunflower three yards across, polished to a mirror-like sheen, towering above the throng," or "the fearsome metallic umbrella of some absent giant, turned on its head."

Napoleon III made a special visit to see this marvel and offered a government subsidy to the man who created it, for the "sun engine," as it was dubbed by the press, could convert sunlight into useful mechanical energy, powering a steam engine, itself the mechanical darling of the industrial age.

The strange-looking apparatus was a solar reflector and boiler invented by an obscure young academic from Tours, Augustin Mouchot. Made of copper sheets coated with burnished silver, the inverted cone-shaped mirror measured 9 feet in diameter at its mouth and had a total reflecting surface of 56 square feet. Its mirrored cone could direct sunbeams at right angles to all sides of a small central boiler located along the axis of the collector. The French journalist Leon Simonin described the boiler's configuration in a popular Parisian magazine:

On the small base of the truncated cone rests a copper cylinder; blackened on the outside, its vertical axis is identical with that of the cone. This cylinder, surrounded as it were by a great collar, terminates in a hemispherical cap, so that it looks like an enormous thimble. . . . This curious apparatus is nothing else but a solar receiver . . . or in other words, a boiler in which water is made to boil by the rays of the sun.

The remarkable machine could generate enough steam to drive a ½-horsepower engine at 80 strokes per minute. When Mouchot put it on display, the reaction was one of stupefied amazement—a motor that ran without fuel, on nothing more than sunbeams! The crowd became apprehensive when the steam pressure rose above 75 pounds per square inch. A nervous bystander exclaimed, "It would have been dangerous to have proceeded further as the whole apparatus might have been blown to pieces." The French were impressed that the boiler could also operate a commercial distiller capable of vaporizing 5 gallons of wine in a minute. It struck observers as bizarre, even magical—a kind of perpetual-motion machine. How else could it turn evanescent light beams into the equivalent of the productive work of human muscles? Even to sophisticated Parisians, this was something akin to alchemy.

The Engineer as Hero

National exhibitions like Napoleon III's Great Exhibition of 1867—showcases for the steam engines, locomotives, weapons, and other miracles of the new age of science—were being held with increasing frequency in France and England as the industrial era came to fever pitch. The advances of technology in previous centuries had begun to snowball. The use of machines to augment the sweat and effort of human beings and animals meant that consumer goods could be manufactured on an unprecedented scale. But mechanization depended on the production of iron, and to make 1 ton of iron took 7 to 10 tons of coal. Coal, in addition to wood, was also in demand as a fuel to power the newly developed steam engines and supply heat for the factories springing up throughout Europe.

The two great catalysts at human beings' disposal as the second half of the nineteenth century began were steam power and the large-scale production of steel. France had joined that small league of countries that could make both. But France had an Achilles' heel that Augustin Mouchot was among the first to discern—France was poor in indigenous sources of energy.

The French Revolution had ended the old aristocratic order and marked the ascendancy of the layer below, the emerging class of educated doers and thinkers, men like Mouchot. And though France as a nation was slow to catch on to the possibilities of industrialism that Britain and Germany were already exploiting, it had produced one of the first men to foresee these possibilities, the Comte de Saint-Simon. In the excitement of the French Revolution, it was Saint-Simon, not an Englishman, who coined the term "industrial revolution," long before it became current in Britain or America.

Saint-Simon popularized the idea of "industrialism," though by this word he meant something different than what it has come to mean. For him, it signified a new kind of society—peaceful, orderly, and rational, in which wealth would be produced by machinery. He attracted followers by his call for a breed of "new men"—builders, engineers, planners—to lead the way. In the new society, he wrote, "the real noblemen will be industrial chiefs and the real priests will be scientists."

It was an exciting vision, for this was the age of the engineer, the age of the inventor as hero. In 1853, the British Royal Navy launched the 90-gun ship HMS *James Watt,* thereby conferring an honor upon an engineer normally reserved for dead admirals and members of the royal family. The stereotype of the lone genius, usually from a humble background, struggling with adversity to become a benefactor of humankind, was gently parodied by Dickens in the 1850s in *Bleak House* and *Little Dorrit*. In the following decade, Samuel Smiles established a pantheon of heroes of the industrial revolution in his volume, *Lives of Engineers and Industrial Biography*. These hagiographies painted their subjects as paragons of self-help, combining mechanical genius, self-discipline, thrift, good manners, and a love of the human race. The book went a long way toward establishing a regard for engineers and scientists as being almost godlike that would last until the First World War. Augustin Mouchot's inventive core was cast in that mold.

Energy Fuels Industrial Growth

By 1860, France had recovered from the economic shambles that the French Revolution and the Napoleonic wars had left in their wake. Though war with Prussia loomed, Napoleon III and his Second Empire had turned France into an economic powerhouse rivaling Great Britain, yet France was at a disadvantage to Great Britain and Germany. She had to import all her coal, while her industrializing rivals were both amply endowed with it. The French decided to pursue an aggressive program to step up domestic coal production. There was plenty of coal in Alsace and Lorraine, though it was hard to mine. The plan worked, and output doubled in two decades, providing resources and power for iron smelters, textile plants, flour mills, and the other new industries that began to appear in France in the second half of the century.

Predictably, not all French thinkers felt secure about the nation's energy supply. In 1860 Mouchot, then a professor of mathematics at the Lycée d'Alençon at Tours, cautioned:

> *One cannot help coming to the conclusion that it would be prudent and wise not to fall asleep regarding this quasi-security. Eventually industry will no longer find in Europe the resources to satisfy its prodigious expansion. . . . Coal will undoubtedly be used up. And what will industry do then?*

Mouchot's answer was, "Reap the rays of the sun."

Augustin Mouchot was an unlikely hero in Napoleon III's pantheon of prominent citizens of the Second Empire. He was a gangly, bespectacled schoolmaster with a shock of black hair and a stringy moustache. His pale, starved visage stares out from the only known photograph of him like that of a frustrated anarchist—he bears an uncanny resemblance to Edgar Allan Poe in Poe's declining years. When nervous, he pulled manicly at the wispy points of his moustache, causing sophisticated friends to laugh. He spoke in a whisper and seemed ill at ease in company.

Mouchot often told his students at Tours that the point of invention from the earliest days of fire, iron, and the wheel was to give human beings some additional advantage over the natural environment. For much of history, the advantages obtained were, on the whole, occasional, on a small scale, and, at least in the early stages, local in effect. Conditions of the natural environment continued to dominate the human condition, which was for the most part (as Thomas Hobbes, the great English pessimist, had said) solitary, poor, nasty, brutish, and short. But now that was changing.

By 1867 science was well on its way in the effort of people to improve the conditions in which they found themselves, to bring the natural environment under control. Human beings were, in effect, replacing the natural environment with a synthetic one, an environment of their creation. Central heating became the vogue, and hot water piped to one's washbasin became possible. These were luxuries unknown to kings in earlier epochs.

A new idea of progress had taken root in the public mind. Progress, vaguely conceived as a rapid improvement in general prosperity and happiness, became a living force. There was a discernible improvement in the outward conditions of life. The person in the street, before whose eyes the marvels of science and invention were constantly displayed, could not help but note the unprecedented increase in wealth, the growth of cities, the new and improved methods of transportation and communication, the greater security from disease and death, and all the conveniences of domestic life unknown to previous generations. Such an observer accepted the doctrine of progress without question. The world was obviously better than it had ever been before. Equally, it would obviously be better tomorrow than it was today. That mindset, in a sense, was the whole point of the technology on display at the Great Exhibition and why average people flocked to see it. Science was an incredible, heaven-sent discipline that could tame the world and make life more comfortable for everyone. The Luddites and others who disagreed were in a minority.

Mouchot's Early Inventions

Augustin Mouchot was 35 when he began his research in the fertile valley of the Loire in 1860. His goal was to find a way to collect the sun's energy in enough quantity to drive industrial steam engines. To Mouchot, the steam engine was technology of the highest order, a device that had revolutionized the France of his childhood. Mouchot understood the principle of the steam engine well. Its enormous power was based on a simple fact: When water was boiled into steam, its volume increased 1600 times, producing a powerful force that could be used to move a piston back and forth in a cylinder. The piston was connected to a crankshaft that converted the back and forth movement into rotary motion; the rotary motion could drive machinery—*any* machinery. The steam engine was a universal machine that produced only one product, but one that was incredibly useful: It *manufactured power*! And that power could supplant human labor as well as the unreliable energy of waterwheels and windmills.

For Mouchot, the magic of the steam engine was that it converted heat, a static (some would say useless) form of energy, into mechanical motion that could do useful work. And because steam was such a powerful force when confined in a boiler, steam engines could harness energy on a huge scale, overcoming hundreds of tons of resistance. This tremendous access to mechanical power was at the heart of the industrial revolution.

Mouchot's first experiments discouraged him. He found that a solar collector large enough to run an industrial machine would take up a great deal of room. It would also be much too expensive to compete with coal boilers. In theory, it was possible to make a solar collector hundreds of feet across, a scaled-up version of the device he eventually demonstrated for Louis-Napoleon in Paris. The technology existed to build such a device—after all, it would be a simpler affair than, say, the construction of the Eiffel Tower, which was already a technical possibility at the time of Mouchot's early work (though it would not rise above the rooftops of Paris until 1889). But such a construction would have been exorbitantly expensive, defeating Mouchot's central hope that solar power would displace coal and wood on a cost basis.

A second design for solar collection provided greater exposure to the sun's rays. Mouchot began to think he might be able to make an inexpensive device after all. The new design was revolutionary; it consisted of a bell-shaped copper cauldron coated on the outside with lampblack, covered by concentric glass bell jars "to retain, as in a trap, the heat of the sun." The spherical design guaranteed that the sun struck some part of the bell jar perpendicularly at all times of the day, despite its movement across the sky.

By contrast, previous models had to be moved by hand to keep them oriented toward the sun. Mouchot found that the apparatus could collect "practically all the rays falling upon the exterior bell, which is to say a rather large sum of heat relative to the volume of the apparatus." Nevertheless, he was still concerned that to run a commercial steam engine, an impracticably large device would be required.

The solution Mouchot tried next combined two solar developments that over previous centuries had evolved independently: the glass heat trap (an idea that would also be adopted by Frank Shuman) and the burning mirror. Mouchot appreciated that a solar reflector, like a child's magnifying glass, could concentrate more sunlight on the collector than the collector could receive on its own by being exposed to the sun.

Mouchot's great insight was to link these two approaches. With the concentrated beam of sunlight, a glass heat trap could be kept to manageable size while still producing enough heat to boil water in large quantities. Mouchot believed that an array of mirrors to concentrate sunlight was "indispensable to making the solar device practical." The new device led to several successful inventions, including a solar oven, a solar still, and a solar pump as well as the giant solar engine that so fascinated Louis-Napoleon and generated the financial support Mouchot needed.

The solar oven consisted of a tall, blackened cylinder of copper surrounded by a cylinder of glass, with a 1-inch air space in between. Food went inside the copper cylinder, which was then covered by a wooden lid. The solar mirror was shaped like a large half-cylinder standing vertically; it faced south and reflected a bright band of sunlight onto the sides of a central cylinder. The mirror was made of polished silver sheets attached to a wooden frame. Mouchot later recounted in his memoirs that he cooked "excellent" dinners in this apparatus:

> *This new oven allowed me, for example, to make a fine pot roast in the sun. This pot roast was made out of a kilogram of beef and an assortment of vegetables. At the end of four hours the whole dinner was perfectly cooked, despite the passage of a few clouds over the sun, and the stew was all the better since the heat had been very steady.*

With a few modifications, Mouchot converted the solar oven into a still that could process wine into brandy. The glass-covered copper cauldron served as the boiler in which wine was heated to a vapor, which was then cooled and collected in a conventional condenser. In his first experiment, Mouchot filled the cauldron with 2 quarts of wine. A few hours later, he reported that he had the pleasure of drinking the first brandy ever distilled by the heat of the sun, remarking that it had a "most pleasant flavor."

Mouchot's solar pump was similar to the basic design of the solar oven and still. A tall, hollow copper cauldron surrounded by two glass covers was soldered on top of a short tank filled with water. A cylindrical reflector concentrated the rays of the sun on the cauldron, heating the air inside. The expanding air exerted pressure on the water in the container below. Within 20 minutes enough pressure had built up to shoot a jet of water through a nozzle attached to the container, producing a spray 10 feet tall that lasted over half an hour. During succeeding attempts, it pumped a continuous stream of water 20 feet up into the air.

Though he had started his solar work only in 1860, Mouchot took out his first patent—for an early solar boiler—No. 48622, on March 4, 1861. He calculated that, on average, 86 square feet of reflecting surface would be needed to generate 1 horsepower. To allow for heat loss, he doubled the area, making it 172 square feet. In his patent application, he correctly noted that the power of the collector varied with the square footage of the reflecting surface, not the surface area of the boiler, where the heat was transmitted to the water. The device he patented had a boiler capacity of 3½ pints of water. It consisted of two cylindrical concentric copper vessels with domed tops and the water space between them. The height of the outer vessel was 16 inches. A glass bell placed at the focus of the conical reflector covered the boiler. The water boiled in 1 hour from an initial temperature of 50 degrees Fahrenheit.

Collecting the Sun's Energy

Though he had made a good beginning, Mouchot had not yet attained his main goal, which was to drive a conventional steam engine with sun power. Solar fountains, stills, and ovens were interesting curiosities at best—at worst, they were toys for the amusement of children. To be taken seriously by the engineers of his day, he knew he would have to be able to generate enough steam to run machinery.

Mouchot already recognized what the main obstacle was. He knew that the large volume of water inside the copper cauldron of his solar heater took a long time to boil, and the device produced steam too slowly and at pressures too low to drive an industrial steam engine. The steam engines of mid-nineteenth century France were leaky. Unless steam pressure from the boiler was continuously high despite the leaks, these engines could not develop enough "head" to operate. The boilers that powered steam engines therefore had to have very large intakes of heat and water in order to make copious and continuous volumes of steam.

To solve this problem, Mouchot substituted a 1-inch-diameter copper tube for the cauldron, increasing the surface area of copper relative to the water to be heated. The smaller volume of water in the tube, he reasoned, would heat much faster, generating steam quickly. To collect the steam, Mouchot soldered a metal tank to the top of the tube. The solar reflector consisted of a parabolic trough-shaped mirror that faced south, tilted to receive maximum solar exposure.

Mouchot reported what happened when he connected this boiler to a specially designed engine that was less leaky than most:

> *In the month of June 1866, I saw it function marvelously after an hour of exposure to the sun. Its success exceeded our expectations, because the same solar receptor was sufficient to run a second machine, which was much larger than the first.*

Mouchot had invented the first steam engine to run on energy from the sun, converting solar energy into useful mechanical work. Over the next 3 years Mouchot continued to refine his design for a solar motor. Its principal weakness, he recognized, was in the configuration of the boiler, which did not produce the volumes of steam that a coal-fired boiler could make with ease.

To increase the boiler's steam-generating capacity so that it could run a large industrial-sized machine, he replaced the copper tube with two bell-shaped copper cauldrons, one inside the other. The double cauldron was then sheathed in glass. The space between the copper shells held a somewhat greater volume of water than had the previous model, but the layer of water was thin enough to heat rapidly, given the large surface of copper available to receive the concentrated solar rays.

Once government engineers had overcome their initial skepticism, the French government, ever conscious of the need to find substitutes for coal at home and overseas, gave Mouchot the financial backing to build an industrial-scale solar engine. To take the design to this next, more powerful level, he built a 7-foot-long cylindrical boiler based on the double-cauldron design with a tall trough-shaped reflector that faced south. Mouchot also added a clock-drive that moved the device from east to west to follow the daily course of the sun, a task previously done manually.

On the whole, the boiler's performance satisfied Mouchot. It vaporized water into steam at a pressure of 45 pounds per square inch, which was at the lower end of the range needed for conventional steam engines, though still not competitive with its coal-fired cousins. To increase the energy input into the device, Mouchot recognized that the mirror design needed further refinement.

Mouchot's analysis of the problem was elegant, leading to an intuitive solution. First, because the mirror's tilt could not be adjusted for seasonal shifts in the sun's path, it did not capture as much of the limited sunlight available at France's northerly latitudes as it should have. Second, the mirror concentrated sunlight only on one side of the boiler—the opposite side remained cooler, lowering the overall efficiency of the machine. And third, as the reflector was built of silver plates and wood, it weighed hundreds of pounds. Mouchot worried that a mirror large enough to run industrial equipment would be too ponderous for the clock mechanism to budge.

Mouchot solved these problems on paper and was preparing to give them form in his workshop when external events intervened. War with Otto von Bismarck's Germany loomed. Copper and other materials Mouchot needed for further experimentation were requisitioned by the military. On July 19, 1870, Napoleon III, goaded by Bismarck, declared war on Prussia. Bismarck's armies were well armed, efficiently organized, and disciplined—and they crushed the once-greater military power of France with ease.

Soon the Germans marched into Paris and the French suffered another ignominious defeat, the more so because it had been unanticipated. Among the victims of the conflict in the capital was Mouchot's largest solar reflector, the one that had earned the admiration of Louis-Napoleon himself—disappearing, as he put it, "in the midst of our national disaster."

A kind of horror descended upon all of France. Nowhere was it more intense than in her capital. Mouchot visited Paris in the winter of 1870, where he found routed French soldiers encamped upon the same Champs-de-Mars where the Great Exhibition had been held such a short time before. The scene of Mouchot's glittering triumph had been reduced to ashes, together with much of central Paris. A string of German victories soon culminated in the French defeat at Sedan, where Louis-Napoleon was deposed, disgraced, and sent into exile. In 3 years he would be dead, the monarchy effectively abolished. Mouchot's wonderful machinery now seemed a dim memory, though hardly an irrelevant one. In a cruel irony not lost on Mouchot, France was forced to give up most of Alsace and Lorraine and her limited supply of coal. At a time when France most needed it, Mouchot's solar technology had been forgotten, crushed in the debris of the conflict.

After the government collapsed and Mouchot's funds were cut off, he tried to find other sources of support for his solar machines. Stuck now in the provinces—he had returned to his beloved Tours in the heart of the rich Loire Valley, south of the cathedral cities of Le Mans and Chartres—Mouchot had a hard time finding money, which he attributed to the "bias and specious objections that engineers pronounced on a question too foreign to their own studies for them to judge." He returned to his teaching in disgust. Science and engineering had given rise to a military evil the world had not seen before. He believed, for a while, that every scientific discovery, even his own, could be perverted. Science had turned from emancipator of humankind to its destroyer.

Mouchot had hoped, with so many thinkers of his day, that scientific knowledge—the mechanism that allowed human control over nature—would foster human self-control through the same rational intelligence that he had brought to his workbench. It was in the name of social improvement that the French aristocracy had been swept away. And it was in the name of social improvement that machines had been welcomed. And yet, France had once again collapsed into chaos. The social struggles among humans had absorbed the struggle of humans with nature and eclipsed it. Men like Krupp had abused the new technology to make larger cannon, not better lives. German and French citizens alike regarded science as increasing their dominion over nature, but they blinded themselves to the fact that by increasing their ability to indulge collective passions, it threatened the destruction of the world. Mouchot could not have guessed that it was a paradox that would endure into the nuclear age.

A Lighter, Stronger Prototype

Four years passed. By the mid-1870s, France had begun to heal, and Mouchot with it. Bored again with his work in the secondary school where he had taught mathematics and physics, he took the advice of a colleague and approached the regional government of Indre-et-Loire in the wine-producing district in which he lived. He still needed money to fund the new machine—lighter, stronger, with a greater collecting surface—he was building in his spare hours in the workshop behind his house. In a letter to an ally in Paris, Mouchot wrote that the solar prototype he showed the officials "so pleased the general counsel that they gave me 1,500 francs on the spot, making available to me the means for completing the construction of a large solar receptor capable of distilling alcohol and producing a sufficient amount of power for mechanical applications."

The conical mirror concentrator was a simple way of making a large collector. The basic geometry was uncomplicated. In cross section, the mirror was made up of straight-line elements inclined at an angle of 45 degrees to the moving solar rays. Each line element acted like a plane mirror, simply folding the beam and directing it to the axis of the cone. Mouchot could adjust the angle of the cone to something other than 45, which simply placed the focal line above or below the lip of the mirror.

The cone-shaped collector was originally important for another reason. In the manufacturing of mirrors for the concentrating mirror system, Mouchot faced the task of shaping the mirror to the desired curve. In the case of a spherical and parabolical mirror, the curvature had to be in two dimensions. A solid material cannot be curved in two dimensions without either a flow of mass within the material or buckling in the surface. Because the cone required the mirror to be curved in one direction only, it could easily be stamped in a machine shop out of flat sheet metal. A glass mirror with curvature in only one direction was even easier to fabricate. And these cones, he discovered, were fantastically effective; the solar flux arriving as the concentrator in a Mouchot device was up to 115 times the normal solar flux.

With his customary alacrity, Mouchot finished a perfected model of his new solar machine in 1874. He put it on public display in Tours, the departmental capital, where it remained for more than a year. In the widely read *Revue des Deux Mondes* Leon Simonin described the machine:

> *The traveler who visits the library of Tours sees in the courtyard in front a strange-looking apparatus directed skyward. It is made of copper, coated on the inside with thin silver leaf. . . . This steam generator is designed to raise water to the boiling point and beyond by means of the solar rays which are thrown upon the cylinder by the silvered inner surface of the conical reflector. The boiler receives water up to two-thirds of its capacity through a feed pipe. A glass tube and a steam gauge communicating with the inside of the generator, and attached to the outside of the reflector, indicate both the level of the water and the pressure of the steam. Finally, there is a safety valve to let off the steam when the pressure is greater than desired. Thus the engine offers all desirable safety and may be provided with all the accessories of a steam boiler.*
>
> *The reflector, which is the main portion of the generator, has a diameter of 8 feet at its large end and 3 feet at its base and is about 3 feet in height, giving about 60 square feet of reflecting surface or of insolation. The interior walls are lined with burnished silver because that metal is the best reflector of heat rays; still, brass with a light coating of silver would also serve the purpose. The inclination of the walls of the apparatus to its axis measures 45 degrees. Even the ancients were aware that this is the best form for this kind of metallic mirror with a linear focus, inasmuch as the incident rays parallel to the axis are reflected perpendicularly to the same and thus give a focus of maximum intensity.*

The boiler is of copper, which of all the common metals is the best conductor of heat; it is blackened on the outside because black possesses the property of absorbing all the heat rays just as white reflects them; and it is enclosed in a glass envelope, glass being the most diathermanous of all bodies; that is to say, the most permeable by the rays of luminous heat. Glass further possesses the property of resisting the exit of these same rays after they have been transformed into dark rays [heat] on the blackened surface of the boiler.

The boiler proper of the Tours solar engine consists of two concentric bells of copper, the larger one, which alone is visible, having the same height as the mirror (about 3 feet), and the smaller or inner one about half that. The feed water lies between the two envelopes, forming an annular envelope an inch thick. The volume of liquid is about 10 gallons and the steam chamber has a capacity of 5 gallons. The inner envelope is empty. Into it pass the steam pipe and the feed pipe of the boiler. To the steam pipe are attached the gauge and the safety valve. The bell glass covering the boiler is about 3 feet high and a foot and a half in diameter. There is everywhere a space of an inch between its walls and those of the boiler, and this space is filled with a layer of very hot air.

On May 8, 1875, a fine day, 10 gallons of water at 70 degrees Fahrenheit was introduced into the boiler at 8:30 A.M. *and produced steam in 40 minutes at 2 atmospheres (30 pounds per square inch)—i.e., at a temperature of 250 degrees Fahrenheit. The steam was then raised rapidly to a pressure of 5 atmospheres (75 pounds per square inch), and if this limit was not exceeded, it was because the sides of the boiler were only ⅛ inch thick, and the total pressure supported by these sides was then 80,000 pounds. It would have been dangerous to proceed further, as the whole apparatus might have been blown to pieces.*

Toward the middle of the same day, with 8 gallons of water in the boiler, the steam at 212 degrees Fahrenheit—that is to say, at a pressure of one atmosphere—rose in less than a quarter of an hour to a pressure of 5 atmospheres, equal to a temperature of 307 degrees Fahrenheit. Finally, on July 22, toward 1 P.M.*, an*

> *exceptionally hot day, the apparatus vaporized 2½ gallons of water per hour, which is equal to a consumption of 70 gallons of steam per minute and generation of one-half horsepower. For these experiments the inventor used an engine which made 80 strokes per minute under a continued pressure of one atmosphere. Later on it was changed for a rotative engine—that is to say, an engine with a revolving cylinder [a turbine]—which worked admirably, putting in motion a pump to raise water, until the pump, which was too weak, was broken.*

By 1878, Mouchot had quintupled the size of the boiler that created the solar-heated steam from a relatively modest 10 gallons to an industrial-grade 50-gallon tank. He invented a boiler design made of many tubes placed side by side with a capacity of 35 gallons for water and 15 for steam. To validate his work, Mouchot appears to have been the only inventor of a solar plant (except Frank Shuman) who had his apparatus tested by independent engineers. M. A. Crova did this in 1882 with the results reported in *Comptes Rendus* that year.

On to Algeria

Even in the sunny Loire, gray days could render the solar engine powerless. Mouchot recognized early on that a significant impediment to the commercial use of solar engines in France was the problem of intermittent sunshine, especially in winter, and the nation's northerly latitude. But France, like Great Britain, had tropical territories more abundantly blessed with sunshine. Mouchot believed that France's baking colonies in northern Africa and Asia, many of which had recently been conquered or annexed and were just being opened up to French settlement, offered unlimited possibilities. As he later recorded in his memoirs,

> *In torrid zones such as Cochin China [South Vietnam], the matter of hygiene comes to the fore. In Saigon, water has to be boiled to be made potable. What a savings in fuel one could realize using a solar still in the ardent heat of those climates!*

The combination of constant sunshine and cheap land convinced Mouchot that solar power would be commercially successful in the French colonies. He decided to make this kind of solar development the focus of his search for government financial aid to develop his machines.

He made an important ally in his search for funds a year after opening his exhibit at Tours when a local aristocrat with connections and investments in North Africa saw a demonstration of the machine. This man, a baron, became a strong advocate of solar power. Through his influence in, improbably, the French Ministry of Public Education, he helped subsidize a scientific expedition to the new French colony of Algeria so that Mouchot could determine whether solar cooking, distilling, and pumping would be practical there.

The French government[1] wanted to spur a second wave of colonial settlement in Algeria to follow the one that had occurred after the insurrection in Paris of 1871. The baron helped Mouchot hitch his project to that star. France was suffering from her painful defeat at the hands of the Germans. In the once-powerful nation, the memories of Louis XIV and the first Napoleon were dimming. If France, under attack in Europe, was to reconquer her position in the world, she would have to look beyond European shores. She would become Africa's great colonial power, controlling by 1911 a quarter of the entire continent.

Although the confiscation of native Berber lands and commercial holdings in Algeria by French interests made it easier for Europeans to acquire property, which attracted shiploads of Frenchmen and their families to the fertile land, the lack of an indigenous fuel supply hindered the colony's economic development—a problem parallel to that the British would face three decades later in Egypt. Algeria had to import all her fuel from Europe, 85 percent of it from France's ancient rival, Great Britain, in the form of expensive coal. The deficiency of railroads in the colony drove up the cost of coal even further—especially in remote districts where prices were

[1]Now in its new incarnation as the Third Republic, a durable government that would last until the Vichy regime of Henri Petain in World War II.

up to 10 times higher than in more accessible regions. The French government hoped that solar power could augment Algeria's economy.

Mouchot arrived in the bustling Mediterranean port of Algiers on March 6, 1877, and set about finding a place to live and work. His goal was a modest one: an improved solar oven for French troops. "It seems possible without any great cost to provide our soldiers in Africa with a small and simple portable solar stove, requiring no fuel for the cooking of food," he wrote back home. "It would be a big help in the sands of the desert as well as the snows of the Atlas Mountains."

Somewhere on the western side of the great melting-pot city, out in the verdant coastal plain separating the Mediterranean from the towering chain of the Atlas, Mouchot rented the crumbling summer villa of a now-penniless Algerian trader. A hundred yards from the banks of an overgrown canal, the mansion was an unwieldy pile of stuccoed stories built by a French architect a few years before.

It was not the ugly building that attracted Mouchot. Behind the villa was a magnificent garden, with forests of peach and apricot trees, almonds, cherries, and plums, filling the garden with their mingled smells and colors—soft clouds of pink and blue against the black-green background of resinous native foliage. Left to itself for several seasons, free from pruning knife and scythe, the garden's alleys had become shut out from the sky beneath a maze of interlacing boughs choked with high grass. There were hedges of geraniums ablaze with scarlet color, hibiscus trees with deep crimson flowers, and wild roses growing everywhere. Groves of palms guarded the gravel alleys.

Mouchot, a child of France's most lush countryside, was entranced. In the center of the overgrown garden stood a huge and rather ugly kiosk or gazebo, circular, much like the bandstand in the center of a park in Tours he was fond of. Mouchot spent the next few weeks ruminating in the cool shade of this covered space, his only company an austere Berber gardener who drank tea under the vines and fig trees, eyeing him warily. There Mouchot drafted the plans for his next solar invention.

At night he returned to the huge villa. The steps outside were already in ruin, some of the stones having been pilfered. The main room was a garish hall set with dozens of tall mirrors in frames of tarnished gold and still bearing the pasted label of their maker in Marseilles. The stone-paved floor was covered with immense French carpets decorated with leviathan roses and lilies. Mouchot found a small room on the second floor in which to sleep. The garish mirrors gave him an idea. He soon dismantled them from their gilt frames for an experiment.

With the help of metal artisans recruited in the narrow alleys of the Casbah where they earned their living hammering copper and brass into teapots and trays, Mouchot built a solar oven with a truncated conical reflector much like the one at Tours, with a glass-enclosed cylindrical metal pot (serving as the boiler) sitting at the focus of the reflector. The apparatus weighed only 30 pounds and

A solar cooker patented by William Adams in India. Adams discovered, like Mouchot, that concentrated direct sunlight on meat creates an offensive odor and taste unlike that of fire-barbecued meat. He adapted red and yellow glass filters to block the ultraviolet rays responsible for this phenomenon, thus adapting the cooker for use by Indian regiments. The Mouchot design was similar.

could be collapsed and packed into a 20-by-20-inch box for easy transport in caravans. With a group of appreciative French colonial officials looking on, Mouchot baked a pound of bread in 45 minutes, over 2 pounds of potatoes in an hour, a beef stew in 3 hours, and a perfect roast ("whose juices fell to the bottom of the pot") in less than half an hour.

With this and other solar cookers Mouchot discovered, as William Adams had before him, that it was necessary to block some of the solar rays to prevent an unpleasant burning odor emanating from the food being cooked. Animal fat, when exposed to the direct or reflected rays of the sun "was converted into butyric acid," Adams had recorded in *Solar Heat,* "a substance having such an offensive odor and taste as to render the most tender roast unpalatable." Mouchot discovered that a sheet of red, pink, or yellow transparent glass interposed between the roast and the reflector had the effect of preventing this chemical reaction. Later experiments would show that it was only the high-energy ultraviolet rays that caused the unpleasant chemical change. Red, yellow, and pink glass all had the effect of blocking these rays, admitting only visible light and infrared heat rays, which allowed the solar oven to function as well as any conventional oven burning wood or gas.

By removing the cooking pot and replacing it with bottles of freshly processed local red wine (which Algeria was now starting to produce in quantity), the oven could be used as a pasteurizer. The sun heated the bottles, killing bacteria that might later multiply when the wine was shipped to France.

Mouchot imagined that Algeria would "be able to ask from the beautiful sun not only to care for the ripening of her vineyards, but also to improve her wines and make them transportable, which would be for her a new source of prosperity."

Mouchot also tested a solar still, similar to the solar machine he had exhibited at Tours. He wrote that the brandy he made from wine was "the subject of astonishment . . . [for] it is undeniable that the alcohol comes out of the solar still bold [and] agreeable to the taste, and with an appropriate wine it offers the savor and bouquet of an aged eau-de-vie." The device could distill fresh water and salt water as well—desalinating water was an important consideration in any desert country. The French Foreign Legion used the ovens and stills for many years. The Paris press reported that "one of the great services that we owe to Mouchot's appliances is the distillation of brine water heavily charged with magnesium salt, which is abundant in the African desert. His still is a great benefit to settlers and explorers."

Mouchot traveled from the Sahara to the Mediterranean to test the feasibility of using solar water pumps for irrigation. In Roman times, the North African coast had been the breadbasket of Europe. Desertification had changed that. Now irrigation was key to agricultural development in Algeria, and coal-powered pumps were proving too expensive to water the land. Mouchot experimented with a solar pump similar to the Tours machine and found that it worked more reliably in Algeria's sunny weather than in France's variable climate, though it was too small to generate a flow of water to make a real difference.

A New Sun Machine

Mouchot, meanwhile, moved out of his isolated villa into the French quarter of Algiers. He was shocked by the contrast between the hovels put up by the French authorities for the *pieds noirs*[2] and the smart villas provided for government officials and managers. All of these were superior to the slums of the indigenous Berbers and Arabs. Here the locals "lived and died without either memory or hope," he wrote, "happy for the crusts that kept them alive or the sleep that brought them the brief, uneasy solace of dreams."

The poor "propagated like flies," reached early maturity—10 or 12 years of age—and promptly went to work. Labor was a resource to be exploited in France's colonies, to be mined, to be exhausted, and finally to be discarded. Years earlier in his classroom at Tours, when not expounding on mathematics, Mouchot had taught his students a reading of history that asserted that shape and meaning were

[2]Literally, "black feet," the term adopted for the poor white migrants who were being sent from Europe, mainly France, to colonize Algeria.

given to society by the adroit maneuvers or blunders of men who appeared in public life as generals, senators, prime ministers, foreign secretaries, kings, and ambassadors. Through his solar experiments, he now achieved a different understanding of history—one that professed that "little men in black coats," the unglamorous technicians who invented steam engines and electric turbines in the back corners of dusty laboratories, could produce astonishing changes. One had only to see an Algerian peasant fleeing in terror at the first sight of a locomotive to understand that something had changed in the world, that the industrial revolution was shaping how people lived, even outside Europe, in ways that politics alone could not. Mouchot recognized in himself one of those little men in black coats.

Mouchot was a citizen of an age when it was difficult to disentangle science from the social and economic factors to which it gave rise. Aware that he was in some sense the instrument of French power in Algeria, Mouchot abandoned his plans for a larger solar machine. He asked for a leave of absence that would permit him to return to Paris. The French colonial authorities, eager to possess the inventions they felt they had already paid for, refused. But they lent Mouchot a sympathetic ear and promised to improve local conditions. Temporarily, he seemed mollified.

Following a year of testing designs for an ever-larger concentrator suitable for tropical use, Mouchot presented his findings to the authorities in Algiers. They were so impressed with the models and sketches that they awarded him 5000 francs to construct "the largest mirror ever built in the world" for a huge sun machine that would represent Algeria in yet another universal exposition in Paris. There, it was hoped, it would garner a number of prizes, win additional financing, and then be shipped back to Africa and used commercially. His money in hand, Mouchot returned to France to begin construction of his new project.

With the help of a new assistant, Abel Pifre, Mouchot completed the new solar machine in September 1878. At its widest point, the cone-shaped mirror measured twice the diameter of the device shown at Tours the previous year, and its total reflecting surface was four times greater. The boiler too had an innovative design: long vertical tubes were fastened side by side to form a circular column at the focus of the reflector. As planned, the solar engine was put on display in the French capital. Mouchot's giant solar machine entertained exposition visitors by pumping 500 gallons of water per hour, distilling alcohol, and cooking food. It was able to achieve a head of steam in far less time than previous models.

The most noteworthy demonstration occurred on September 22, 1878, as Mouchot recounted: "Under a slightly veiled but continually shining sun, I was able to raise the pressure in the boiler to 91 pounds ... [and] in spite of the seeming paradox of the statement, [it was] possible to use the rays of the sun to make ice."

Mouchot was able to make ice because he had connected the solar motor to a heat-powered refrigeration device invented by

A Mouchot solar concentrator (right), boils water to run the small steam engine (center) which runs Abel Pifre's solar-powered printing press (left). Pifre, Mouchot's last assistant, ran off 500 copies of the *Journal Soleil* at this exhibition in the Tuileries gardens in Paris in 1880. The gearing beneath the concentrator allowed an operator to point the cone directly toward the sun and follow its path through the sky by turning a hand crank. Note also the smaller Mouchot cookers and concentrators in the foreground.

Ferdinand Carré in the 1850s. Mouchot saw an important future for solar refrigerators in hot climates, where sun-generated ice would help prevent food from spoiling. The average Parisian spectator, hardly aware of the scientific principles at work, was amazed—this was magic indeed, the burning heat of the sun transformed into ice.

Storing Solar Heat

The following year Mouchot returned to Algeria to resume his research, intrigued now by a new question that had perplexed him in the past but that he had put off: how to run solar machines when the sun was not shining. He spent much of his time trying to resolve this difficulty. Could solar heat be stored so that sun machines could work during cloudy weather or at night? A colleague suggested using heat-absorbing material—water, for example—capable of withstanding the high temperatures produced by a solar reflector. By placing the heated material in an insulated container, solar heat could be retained for later use. But Mouchot was not impressed with this idea because heated water (or any heated substance) lost heat to its surroundings, even with the best available insulation.

Mouchot discovered what he thought was a better alternative. If solar energy were used to break down water into hydrogen and oxygen, the gases could be stored in separate cylinders. When heat was needed, the chemical reaction resulting from the recombination of the two gases would produce high temperatures—a flame. Or the gases could be used separately—the hydrogen as fuel and the oxygen for industrial purposes. As for the method of separating water into its components, Mouchot decided to try "an instrument already in excellent condition . . . the thermoelectric device."

The principle behind it was simple and well understood by the science of Mouchot's day. When two different metals such as copper and iron are soldered together and heat is applied to the juncture between them, an electrical current results. Mouchot planned to heat a hundred such metallic couplings with a solar reflector and thereby generate enough electricity to change water into its constituents.

By 1879, Mouchot had "already made a few experiments which bode well for this procedure.... Some primitive devices have given me significant amounts of electricity," he wrote. Mouchot had great expectations. He hoped to decompose water to make "a reserve of fuel that would be as precious as it is abundant." But for all his efforts, he could not compete with the more efficient methods of electrical generation being perfected about the same time. In 1880, depressed by the latest round of failures, Mouchot returned to his mathematical studies. His gifted assistant, Abel Pifre, took over the solar research. Pifre built several sun motors and conducted public demonstrations to renew public support for solar power. Some were little more than publicity stunts. At the Tuileries gardens in Paris he exhibited a solar generator that drove a press, as reported by the British journal *Nature*:

> *There was set up at the Jardin des Tuileries at Paris, near the large reservoir at the foot of the Jeu de Paume stairs, an insolator that actuated a Marinoni printing press. Although the sun was not very hot and its radiation was interfered with by frequent clouds, the press was able to work with regularity between one and five o'clock in the afternoon and to print on an average five hundred copies per hour of a journal specially composed for the occasion, entitled* Journal Soleil.

But the time wasn't right for solar energy in France. The advent of better coal mining techniques and an improved railroad system (most of France's coal lay at her borders) increased coal production and reduced fuel prices. In 1881, the government took a final look at the potential for commercial use of solar energy. It sponsored a year-long test of two solar motors—one designed by Mouchot and the other by Pifre. The report concluded:

> *In our temperate climate, the sun does not shine continuously enough to be able to use these devices practically. In hot and dry climates, the possibility of their use depends on the difficulty of*

Another view of a Mouchot concentrator being used to power Abel Pifre's printing press. This illustration appeared in Britain's *Nature* in 1882 with the caption: "The *Times Corrrespondent* in Paris mentions having seen, at a recent popular fête at the Tuileries, a solar apparatus set in motion a printing machine, which printed several thousand copies of a specimen newspaper called the *Journal Soleil*. He also saw cider and coffee made with its aid, and a pump set in motion. He suggests the use of such appparatus for troops in Egypt and India."

obtaining fuel and the cost and ease of transporting these solar devices.

Furthermore, the cost of constructing silver-plated mirrors and keeping them polished proved economically prohibitive for most uses. During ensuing years the French Foreign Legion made some use of solar ovens in Africa. In remote areas of Algeria, people also used solar stills to obtain potable water.

Although Mouchot in the end did not succeed in converting France to solar power any more than Napoleon III was able to create a newly invigorated French empire, his pioneering work crossed the threshold between scientific experimentation and the practical development of a revolutionary technology. Augustin Mouchot demonstrated, for the first time and in a way that could not be discounted by his critics, that solar energy could be used as a source of mechanical energy, not merely heat and light. He had laid the foundation for future solar developments that would fill scientific journals and the popular press for the next 150 years.

Mouchot, in his own mind, hardly considered this to be enough. He was at heart a social activist. He felt that to conquer nature, one had also to conquer the nature of human beings. He thought technology had taught the French to become gods before they had learned to be humane. The struggle to harness solar energy had become subsumed for Mouchot in the social struggles of Europe because the instruments through which people transformed the resources of nature—steam engines, locomotives, looms—into means for the satisfaction of desires were regarded more and more as the objects of political conflict.

It turned out that was an insight more welcome in British intellectual circles than in the minds of Mouchot's fellow French citizens. In British-held Egypt a third of a century after Mouchot had left Algeria, Frank Shuman would find a more receptive political and social environment for solar power, and he would focus his efforts exclusively on the technical challenges of the large project he had committed to build.

7
Egypt's Great Sun Machine

When the solar-powered irrigation plant opened at Maadi, Frank Shuman brought all his considerable skills as a showman to bear on the event. He invited every journalist, senior civil servant, and diplomat in Cairo and Alexandria to visit the site for a grand kickoff festival. Joseph Callanan describes the celebration in *The Great Sun Machine* (an account published by Exxon in 1975):

> *The scene has a touch of post-Edwardian elegance. The time is 1913, and there is a party on the banks of the Nile. The guests, mostly British, move dreamily among the palm trees. Ladies carry parasols and gentlemen wear topees against the tropic sun. The thermometer tops 100 degrees Fahrenheit. There is cold champagne and the heat is not entirely unwelcome. The burning sun, in fact, figures prominently in the occasion.*

For the grand opening of the plant, the guests of Frank Shuman and A. S. E. Ackermann had received engraved invitations from the Sun Power Company (Eastern Hemisphere) Ltd. printed by one of the tonier Cairo printers on their best pasteboard stock. Waiters bearing trays of tiny sandwiches and slivers of bread and toast topped with cheese and caviar moved unobtrusively among the guests. For the thirsty, there were bottles at the ready to refresh empty glasses with generous servings of Veuve Cliquot; for the teetotalers, there was lemonade. The German consul general was there with his whole entourage—a group, it was noted, that paid the most dedicated, minute attention to every detail of the machinery on display, snapping scores of "Kodaks" of the plant from all conceivable angles. The machinery was impressive. Callanan writes:

> *The steam engine that drives the pump is fueled by the sun's rays, nothing else. The guests hear the engine hiss into action and they see Nile water gush over the land. They are told that the sun engine generates 65 horsepower and pumps 6,000 gallons per minute. This is most impressive—the sun is free, and coal here costs many pounds sterling a ton.*

A Sun Power Company hired hand filmed the operation of the sun absorber and machinery using a primitive hand-cranked motion picture camera. Shuman thought he might use the new "movies"

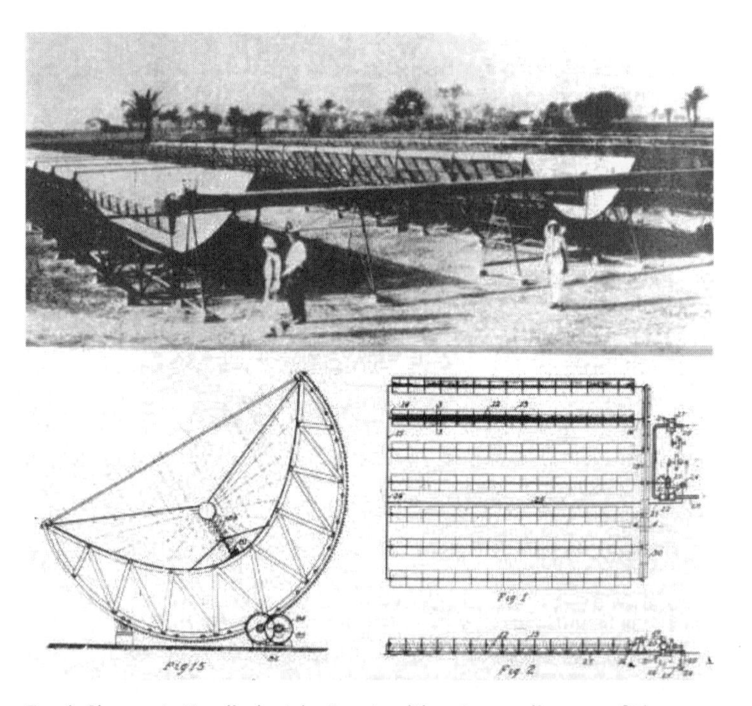

Frank Shuman's Maadi plant in Egypt, with cutaway diagram of the parabolic trough collectors.

technology as a promotional tool, and he would soon enough have that chance. The photographer panned across the dais and focused on the visage of "a leathery man of lordly bearing," in Callanan's words, who appeared to be the guest of honor. This was Viscount Kitchener of Khartoum. He and Shuman were seen to be engaged in rapt discussion as Shuman escorted him around the grounds, pointing and gesturing as he gave the British consul general a personal tour of the solar-powered equipment.

The American inventor and his sun-driven irrigation pump fascinated Kitchener, long an amateur engineer. He saw vast potential for solar-powered irrigation, especially farther inland in the Sudan where coal was even more expensive than at Cairo. Kitchener told Shuman that he was searching for a way to conquer Sudan economically as he had conquered that land militarily a decade before. Could the giant pumps at Qom Ombo and others like them be powered by Shuman absorbers instead of precious imported fuel? It was a question that seized Kitchener's imagination no less than Frank Shuman's.

Shuman explained that his enterprise had overcome the two main obstacles that had foiled others in harnessing the sun's energy. The sun was not constantly available, Shuman observed, even in Egypt, and Shuman himself and other inventors had spent time and ingenuity looking for a means of energy storage to sustain solar-powered systems at night and during periods when local weather conditions obscured the sun. In Shuman's now abandoned 1000-horsepower design, a huge storage tank had figured prominently in the hot water cycle to permit the machinery to run 24 hours a day. In a plant used exclusively for irrigation, as in Egypt, heat storage was not needed, Shuman explained, and its cost was one that could be avoided entirely. It was necessary to operate the plant only during the day, and cloudy days wouldn't normally make that much difference in getting the required amount of water to the crops. So the problem of storage had been set aside.

The second obstacle that had stymied other inventors was the fact that solar energy was diffuse. Although the total amount of energy from the sun's rays available on any given plot of land in Egypt might be enormous, the collection and conversion of sunlight to do useful work entailed a large capital investment to concentrate the energy. Shuman explained that he believed he had found the best method for collecting the sun's heat in an economically feasible way. Sun-powered irrigation would raise the value of Egyptian wasteland by making the rich soil moist enough to bear crops.

Kitchener was smitten by Shuman's enthusiasm, just as he was exhilarated by the concept of solar power as an economic driver of agriculture. He invited Shuman to visit him at the British Agency in the coming weeks. There they would share meals together and continue their discussion of how Egyptian deserts might be transformed. Soon Kitchener would invite Shuman to build a much larger solar plant, and he would introduce him to Sir Reginald Wingate, governor of the Sudan, where they would plan the largest solar plant of all.

Problems with the Egyptian Plant

Despite the aura of success that pervaded the official opening of the Maadi plant and the opportunity it presented for Shuman's powerful brand of personal salesmanship, the Egyptian venture had been plagued with problems for months. Because of time constraints and the shifts in design recommended by C. V. Boys, the Sun Power Company had abandoned the idea of trying to prefabricate components and assemble them in Egypt. Instead, the team in Maadi worked on the fly, constructing the installation from scratch at the site and making last-minute changes on the spot. The first absorber fashioned along the new Boys configuration was built at Maadi in 1912. This strategy, accompanied as it was by a dearth of planning and forethought, proved both expensive and time consuming.

The original tubes for the boiler were ordered to be made from zinc, an elemental metal that was ductile and malleable when heated, so much so that the boilers buckled in tests and "went limp as spaghetti," in Ackermann's words. New pipes were ordered, this time made of cast iron. They were installed in early 1913 after months of costly delay.

Shuman also ordered a 100-horsepower, low-pressure reciprocating engine to be shipped across the ocean from his Tacony labo-

Frank Shuman's Maadi parabolic troughs, close up.

ratory. This mix of new and old equipment sometimes ran smoothly, but minor hitches and problems continued to plague the assembly. The pair of engineers conducted a series of trials—35 in all—before they were satisfied the prototype plant was functioning as it was supposed to. Only then were they ready to invite Egypt's elite for the grand opening.

Several innovations occurred to the engineering team during construction and were implemented at the last minute. The axes of the parabolic troughs were oriented north and south, and thermocouples were used to cause them to heel over automatically from east to west as the sun followed its course. This meant that the amount of solar energy captured was steady all day long, with a small decrease in steam production in the morning and evening owing to the greater thickness of the atmosphere through which the solar rays had to pass. The total area of the collectors exposed to sunshine was 13,269 square feet. The boilers were placed at the focus of the reflectors and were covered with a single layer of glass enclosing an air space around the boilers. Each channel-shaped reflector and its boiler was 205 feet long, and there were five such sections placed side by side. The concentration ratio provided by the reflectors was 4½ to 1.

Tests conducted by Ackermann in August 1913 revealed that the absorber could deliver 1440 pounds of steam per hour at atmospheric pressure. According to Ackermann's technical report:

The maximum quantity of steam produced was 12 pounds per 100 square feet of sunshine, equivalent to 183 square feet per brake horsepower[1] *and the maximum thermal efficiency was 40.1 per cent. The maximum output of the engine was 55.5 brake horsepower—a result about 10 times as large as anything previously attained, equal to 63 brake horsepower per acre of land occupied by the plant. A pleasing result was that the output did not fall off much in the morning and evening. Thus on August 22, 1913, the average power for the five hours' run was not less than 59.4 brake horsepower per acre, while the maximum and minimum power on that day were 63 and 52.4 brake horsepower per acre, respectively.*

These results suggested that Shuman's original design for a 1000-horsepower turbine had underestimated by as much as 25 percent the size of the absorber required to run a steam turbine of that size. Extrapolating from Ackermann's field tests, an area 400 feet square (160,000 square feet of asphalt) would have supported a 750-horsepower reciprocating steam engine, not a 1000-horsepower machine.

Comparing the energy needs of a reciprocating steam engine with those of a steam turbine was a tricky exercise at best, because turbines were generally more efficient than reciprocating steam engines, particularly as they got larger. Shuman's original design might have come close to the mark. Had Shuman's partners opted for the larger prototype, it appeared likely to Ackermann that it would have been successful. Also, heat losses in the Maadi plant, which delivered only

---

[1]Brake horsepower, sometimes called shaft horsepower, is the amount of power available from an engine that can be delivered to attached machinery to do useful work. It is always less than the total horsepower produced by the engine because some of the engine's energy is used internally by the engine itself, while some is lost to friction in delivering power to the attached machinery.

50 to 60 horsepower, were vastly larger on a percentage basis than those that would have taken place in a larger installation.

Shuman had learned over the years that one of the biggest problems the designer of solar-powered machinery faced was the problem of insulation and heat loss—how to retain the maximum fraction of heat possible coming down from the sun to perform useful work. No part of sunlight does not contain energy, so the first goal was to make the absorber as nonreflective as possible. It was important that no light rays escape conversion into heat. Once the solar energy had been transferred to steam or hot water in the boiler, the aim was to insulate everything touched by the water to keep the heat from seeping away into the surroundings.

Other things being equal, Shuman had learned that the larger the steam engine, the pipes, and storage tanks, the more efficient they were in preventing heat loss. It was always desirable to make the steam engine as large as practicable and to assemble at one point all the tasks requiring the engine's power. In Egyptian irrigation this meant, in practice, favoring big, centrally located irrigation pumps such as those at Qom Ombo, rather than a larger number of smaller installations like the demonstration project at Maadi.

As early as Kitchener's visit to the grand opening, Shuman had perceived that there was a likely confluence between Lord Kitchener's need for massive irrigation pumping plants and the kind of plant the physics of solar power most favored. He might yet have his chance to produce a solar-powered plant design along the lines he personally favored—a big plant with acres of low-technology heat absorption.

Egypt Offers New Opportunities

Soon after the grand opening of the Maadi plant, Shuman paid a visit to the British Agency in Kas-el-Doubara Street. Following the proper etiquette of the times, he sent in his card with the attaché who greeted him at the door. While the attaché went to find Kitchener, Shuman was led down a hallway by an Egyptian guard. At the end of the long hall, he was shown into a parlor and small private dining room that was part of Kitchener's personal quarters. From the

French windows, Shuman looked across a beautifully laid-out garden with beds of roses and carnations. The lawn went right down to the banks of the river. At the far end of the garden, beyond a low wall hung with vines, was the broad expanse of the glittering Nile itself. Shuman could see the white sails of feluccas gliding up and down the calm waters, just as they had done for five thousand years.

Kitchener soon joined him. He was a tall, strongly built man wearing spectacles (pince-nez, like Shuman) and smoking a cigar. He took Shuman outside into the garden to show him his carnations, which he tended himself and of which he was proud. An early riser, Kitchener had already reviewed the outlines of the plan that he wished to put to the Sun Power Company. The two men talked of science and engineering. Shuman stayed through lunch. In the afternoon, Kitchener took Shuman for a ride in his car round Gezira, then showed him the bazaars where he spent hours searching for the special blue china he had collected for 30 years, which he kept on display at his residence.

Kitchener laid out his plan: He proposed that the Egyptian government finance Shuman's next irrigation plant, this one a behemoth even larger than the facilities at Qom Ombo, to be built at a site on the Nile far upriver in Upper Egypt. Kitchener had studied the blueprints of the Maadi plant and was completely conversant with the technical problems that still needed attention. Shuman listened quietly, nodding his assent. For Frank Shuman, the bold risk-taker and innovator, this was the moment he had been waiting for, the big endorsement that could change everything.

Kitchener's stature in Britain at the time was such that, after the king, no one was better known or more revered. Kitchener's backing would mean a qualitative change in how Sun Power Company would be perceived and in the quantity of resources that would now be made available for new projects. Shuman had worked hard for just this kind of opportunity. They shook hands. It was a deal. Who now could fail to take him seriously?

For the next few weeks Shuman tackled problems directly related to irrigation in Upper Egypt and the Sudan. If Kitchener was serious about finding a way to water vast new acreage in the Sudan, where was the fresh water to come from? Surely the Nile's bounty was already committed, and to use a substantial portion of the Nile's water in Sudan would be to rob Egypt of an equivalent share?

In fact, a huge untapped supply of fresh water was already available, though getting at it would take some doing, for much of the White Nile's charge of fresh water was being lost in the swamps in a mysterious region known as the As Sudd, or simply the "Sudd."[2]

Lake Victoria, a vast inland sea, the second largest freshwater body in the world (after Lake Superior), straddled Tanganyika and Uganda in British East Africa. It was 250 miles long, 200 miles wide, 250 feet deep, and covered some 27,000 square miles (an area about the size of South Carolina). Its only outlet was the Nile. The crystal-clear waters of this inland ocean escaped over Ripon Falls on the northern side of the lake 30 miles north of the equator near Jinja. This huge outflow of water—nearly a trillion cubic feet a year—gave rise to the Victoria Nile, which flowed northwest into Lake Albert only to emerge again a few miles farther along the shore as the Bahr al-Jabal (in Arabic, "River of the Mountain"). This river was known, in its lower and more tranquil reaches, as the White Nile—the most stable and steady source of water for the Nile that would eventually flow across the Egyptian deserts (though it contributed only a small fraction—no more than 10 to 15 percent—of the total downstream volume of the river).

At the time of Shuman's experiments in Cairo, the northern shore of Lake Albert was part of the territory of Anglo-Egyptian Sudan. The headwaters of the Nile in that direction were therefore already under British control. From its exit from Lake Albert for a distance of about 500 miles from the Uganda border to the city of Malakal in southern Sudan, the river passed through a region that, while British territory, was virtually unexplored. This was the vast, flat swampland and rain forest of the Sudd. Here marshes and huge floating islands of water hyacinth and papyrus, along with masses of

[2]The word is derived from the Arabic for "barrier" or "obstacle."

dead vegetation, spread the river water out in shallow pans that stretched for dozens of miles in all directions. Much of the water that flowed through the Sudd simply evaporated or sank into the earth to supply underground aquifers. The Sudd was a snarl of lakes, channels, and swamps that covered from 6500 square miles in the winter dry season to perhaps twice that (an area larger than Belgium) during the summer rains from July to September.

Above Gondokoro, the capital of Equatoria Province and an infamous market center for ivory and slave trading until it was occupied by the British in 1898, the mighty river flowed out of this tangle of swamps just as though it were emerging from a hidden opening in the earth itself. Once at Gondokoro, the upstream terminus of the regular service of river steamers from Khartoum, the White Nile kept once again to its narrow banks and flowed steadily north to be met at Khartoum by the waters from Ethiopia's Blue Nile. The combined rivers then flowed unimpeded to the Mediterranean Sea, 1600 miles downstream.

Though exact figures had yet to be assembled by Kitchener's hydrographers, it was well known in the British Agency in Cairo that the Sudd was the great natural thief of Egypt's fresh water. In round numbers, the White Nile's journey through the 16 degrees of latitude that comprised the swamp robbed half the volume it held when it flowed out of Lake Albert, owing to tremendous losses from evaporation in the huge lagoons and swamps, as well as seepage into the underground water table.

The stark fact was that while Lake Victoria formed the first and largest natural reservoir of the Nile system, much of that precious water simply never made it through to the lower parts of the river. British scientists had already determined that Victoria's heavy rainfall was almost balanced by surface evaporation. Even so, 815 billion cubic feet of fresh water flowed out of the lake and over Ripon Falls every year. Additional fresh water, mainly from the Semliki, augmented the volume of the upper river, so that the annual outflow from Lake Albert was about 920 billion cubic feet a year. The seasonal tributaries of the Al-Jabal supplied it with nearly 20 percent of its water, or another 190 billion cubic feet a year, bringing the total to something like 1.1 trillion cubic feet.

Then the Sudd took its enormous toll. The outflow from the Al-Jabal varied very little throughout the year because of the regulatory effect of its swamps and lagoons, which stole something like half the inflow—about 500 billion cubic feet of water a year (a volume equal to about 3.5 cubic miles of fresh water annually). This then was the volume of water that could potentially be recovered for irrigation purposes in Sudan and Upper Egypt without "stealing" water that was already committed to projects lower down the river. In Shuman's mind, solar pumps could spread out water from the Sudd across millions of acres of fertile desert, turning them into a tropical garden.

Kitchener had a problem. He needed to succeed in making the case for irrigating vast new stretches of desert in the upper reaches of the river with solar power without alienating politicians in Cairo and London. But how to demonstrate that the water lost in the Sudd could somehow be returned to the river? British engineers had been looking at proposals to accomplish this for nearly 20 years. At the time of the Maadi solar demonstration project, Kitchener and Sir Reginald Wingate had ordered more than a dozen studies to investigate ways of providing a greater supply of water to Egypt and the Sudan, illustrating almost all the possible ways of reengineering the world's longest river. The low Aswan dam, already built, stored some of the floodwater from the Ethiopian highlands, but not nearly enough. (In the 1960s the Aswan High Dam would rectify this problem, creating 40 cubic miles of additional water storage capacity.)

Kitchener reviewed and rejected unsatisfactory proposals to continue building small dams higher and higher up the river. These were dropped because dams could not safely be built except on, and into, solid rock, and the Nubian sandstone was so permeable that a dam bedded across the Nile Valley would ooze like water percolating through a beach. The harder rocks that comprised the bed for the cataracts above Aswan were comparatively soft and "rotten" in patches and also unsuited for the foundations of a dam. There were other proposals on Kitchener's desk to build dams above Khartoum and to create an artificial lake in the desert near Wadi Rayan, but these too seemed to fire neither his imagination nor enthusiasm.

Kitchener was also considering building a low dam farther upstream to contain the outflow of Lake Albert. The area of the Albert Nyanza was so enormous (nearly 2000 square miles) that a dam only a few feet high would have held back a huge volume of water. This alone might not have been of much use (since Lake Albert was above the Sudd) but it demonstrated that water management was possible even in the highest reaches of the river and its tributaries. That left open the remaining outstanding problem of finding some way to turn off the spigot of water that flooded the swamps, where it was lost forever.

Long before Kitchener's tour of duty in Egypt, while he was serving as governor-general of the Sudan, British engineers had labored to clear a navigable channel through the Sudd. It had proved to be a near-impossible task. Mountains of vegetation came adrift and fouled the dredges in the channel. Smaller vessels, acting as tenders, were snared and sometimes immobilized for weeks. For 10 years there was little progress. True, resistance to the flow of the river had been lessened measurably and the discharge of the White Nile was increased at Khartoum, but even when the channel was kept as clear as possible, permitting access to paddle wheelers, roughly half the available flow through the vast swamp was dissipated.

A British hydrologist had proposed as early as 1899 that a massive reclamation project take advantage of the Sudd's geography to create a canal that would cut right through the swamp. Instead of meandering 500 miles, the White Nile's water would flow only 150 miles through a ruler-straight, human-made channel, cutting off 350 miles of the river. The swiftly moving, unobstructed stream would be held within the packed-earth banks of the canal, shutting off the Sudd's ability to slow the water down and divert it. Early estimates of the improved water flow showed that the canal would be so effective that it might cause *too much* water to flow through southern Sudan, more than could be handled downriver in Egypt during the annual flood. The ability to divert some of the water into the swamp, as a flood-control precaution, was therefore retained in the plans for the canal.

At those times when ample water was available in the rainy season, the canal engineers would measure out just the amount of water needed to pass along to Khartoum, and the rest would be cut off and discharged into the Sudd as before, diverting any surplus that might endanger the valley below. Even so, Kitchener was confident that, with time, productive use could be found for all the water. This would give the upper Nile valley the potential for at least 500 billion cubic feet of "new water" annually, without robbing Egypt of a drop.

In the waning months of 1913, Shuman reviewed these proposals and became convinced that massive, solar-powered irrigation schemes in Sudan and Upper Egypt were feasible. Through Kitchener's intercession he obtained an invitation from Sir Reginald Wingate, the British governor-general of Sudan and a Kitchener protégé, to visit Khartoum to discuss a proposal. Within weeks, Wingate offered Sun Power Company use of a 30,000-acre cotton plantation (nearly 50 square miles) in British Sudan on which to test a much larger solar-powered irrigation facility.

Germany's Interest in Sun Technology

Meanwhile, by the end of 1913 rumors of war were circulating in Cairo. This was no surprise to the British agent. Kitchener had long seen problems with Germany on the horizon. On a tour of the Middle East in 1910, before his appointment as consul general, he spent a week in Constantinople, capital of the Turkish Empire and the city where he hoped some day to be appointed British ambassador. Writing to a friend in England, Kitchener deplored the influence of the Germans within the Sublime Porte. "We are out of it altogether," he wrote, "as the [British] ambassador does nothing and the German is allowed to do as he likes." Kitchener believed that, had he secured the ambassadorial post in Turkey, he might have retrieved the British position, but by 1913 it was too late. British influence in Turkey had all but evaporated. By the time of his arrival in Cairo as British agent, war had already broken out between Italy and Turkey. Kitchener kept Egypt neutral in that conflict, though public opinion in Muslim Egypt was very much on the side of the Turks.

The German consulate general in Cairo and the consulate in Alexandria were large, well-staffed operations that kept Berlin well informed of political events in Egypt and the region. Africa and the Middle East were important theaters for Anglo-German and Franco-German rivalry, and Germany was eager to increase her share of the territorial pie in the Middle East and Africa. In Iraq, the Germans were battling the British for economic control by building railways and threatening British shipping interests on the Euphrates. The story was much the same in North Africa on the Nile. In East Africa, German interests were busy carving out a German East Africa from Tanganyika to compete with the British enclaves of Kenya and Uganda. Reports of Sun Power Company's demonstration project at Maadi, to the opening of which German officials had been invited, had not passed unnoticed in the German capital. The German consul general had attended the festivities to see the machine for himself. His aides had drafted a detailed report on the construction and operation of the plant for the German Foreign Office.

Throughout the nineteenth century the German government had prided itself on being on the cutting edge of emerging technologies, and this new sun technology, it was clear, could have an enormous impact in Germany's growing colonial possessions. Moreover, Frank Shuman, the German consulate noted, was an American citizen of German extraction with no long-standing ties to Great Britain. Surely he would not fail to accept an invitation to the fatherland to discuss his ideas.

Shuman, a man whose personal sensibilities were all but opaque to politics, in the late summer of 1913 was offered a brief, all-expenses-paid trip to Berlin to address the German Reichstag on the potential for solar power in the modern world. The German consul general in Cairo made it clear that Shuman would be treated regally. No foreign inventor, he took pains to note in extending the invitation, had ever been accorded this honor by the German nation.

Shuman's cautionary business instincts appear not to have been alarmed. On the contrary, he was elated at yet another sign of the growing recognition he was receiving from the highest levels in Europe. He accepted immediately and the trip was soon set for late October. He would sail directly from Alexandria to Hamburg on a German-Amerika vessel and then make the 200-mile journey to Berlin on a special VIP train in the company of a large number of his German hosts.

From the standpoint of British and American prewar diplomacy, the timing of Shuman's trip could not have been worse. On the night of September 29, 1913, Dr. Rudolf Diesel, inventor of the diesel engine, was crossing the English Channel on the steamer *Dresden*. He was on his way to London to attend a meeting of diesel-engine manufacturers and to confer with the British admiralty concerning the use of his motor in naval vessels. At about 10:30 P.M. Diesel said goodnight to two colleagues and went to his stateroom. The next morning when the boat docked at Harwich, he did not appear for breakfast and a steward was sent to rouse him. His berth was turned down but had not been slept in. His nightclothes lay folded on his pillow. His watch had been carefully hung in such a way that it could be seen from the berth, but Dr. Diesel was not there and he was never seen again, dead or alive.

Though best known for the pressure-ignited, oil-burning heat engine that bears his name, a device that had already begun to supplant the large, expensive, and fuel-wasting steam engines of the industrial revolution, the multimillionaire Diesel was also a connoisseur of the arts, a gifted linguist, a social theorist, and the frequent object of press coverage in the society pages of German and British newspapers. His disappearance caused a sensation. As the days passed and neither he nor his body were found, his loss became an international mystery that focused public attention on the role technology might play in the coming conflict in Europe. There were rumors in London that Diesel had been abducted or killed by German agents to keep him from giving technical secrets to the British. (One paper even printed a story alleging that he was carrying secret plans for a new engine that ran on nothing but distilled water.) There were hints of financial troubles and reports he had run away and changed his name. Some said he was depressed over a love affair and had committed suicide—but, if so, where was the body? The English Channel, though swept night and day by treacherous currents, was one of the most heavily traveled maritime corridors in the world. The mystery was never solved.

Germany's technical prowess had long been of concern to British and American intelligence agencies. Within this context, the announcement of a trip to Berlin by an American engineer with a professed interest in, and acknowledged expertise in, energy technology raised concerns. The U.S. Navy Department also noted with alarm that Shuman, among his many other inventions, had designed an apparatus to enable submarines to stay submerged longer than previously possible using a process involving liquefied air. But war was still months away and no official bar was raised to the trip, either by the United States or Great Britain.

Shuman landed at Hamburg in early November according to plan. As promised, the German government had convened a special technical session of the Reichstag to hear Shuman speak. The American inventor was escorted to Berlin in an express train accompanied by an imperial delegation representing the kaiser. In the great well of the German parliament, enunciating in the halting German he had learned as a boy, Shuman described the fantastic possibilities of solar power to the assembled politicians. He even showed them moving pictures of the Maadi plant in operation. Impressed by his presentation, representatives of the German government offered the equivalent of $200,000 in Deutsch marks for a sun plant in German Southwest Africa—for Shuman a huge sum, nearly double his investment in Maadi. This kind of recognition was heady stuff. Shuman accepted the offer with alacrity.

With war approaching, Germany was accelerating plans to enlarge and consolidate her four African protectorates of Togoland and the Cameroon, German East Africa, and German Southwest Africa, the latter being the area we know today as Namibia, on the continent's southwestern coast. Of these four African colonies, Germany's most ambitious hopes by far were pinned to the territories known in the fatherland as *Deutsch Südwestafrika* (German Southwest Africa) for several reasons, not the least of which was that this was the colony that might provide the springboard from which to launch a lightning campaign for the biggest prize of all—the whole of southern Africa, including Great Britain's possessions there.

The view from Berlin prior to the First World War was that the great sunny district of South Africa was a zone perfectly fitted by nature for German expansion. The temperate climate, the presence of the great diamond mines and huge deposits of gold, the continued success of a booming agricultural and cattle-raising sector—all these offered an irresistible prospect for the evolution of a great African dominion for Germany. It was a colony that had the potential to become as large as the fatherland itself and a territory which, in the meantime, could provide a new market for German commerce, a new home for German expatriates who felt crowded by Germany's borders, and a rich new source of agriculture and industry for the growing German Empire.

Moreover, under certain favorable circumstances that now seemed increasingly likely to unfold, planners in Berlin thought that the military cost of this acquisition might be quite low. Strained relations between Britain and the Boers made it improbable that German troops would be required in large numbers to annex South Africa—the Boers themselves would take to the field in Germany's name to rob Great Britain of her valuable colony and hand it over to Germany as a gift.

In the days after his warm reception by the Reichstag, Shuman was approached by the German secretary of state, Dr. Wilhelm Heinrich Solf, who had just returned from an extended tour of Germany's African colonies. Solf wanted Shuman to accompany him the following spring on a reconnaissance trip to Windhoek, in German Southwest Africa, to assess potential sites for the demonstration project the government had just agreed to fund—an installation that promised to look something like the Maadi project, though considerably scaled up. Buoyed by his heroic welcome in Berlin, Shuman enthusiastically recommended to Solf an even larger plant based on the 1000-horsepower, 4-acre absorber concept he had originally proposed for Egypt.

Solf had had wide diplomatic service outside Germany. At the time Shuman met him, he had just completed a 10-year tour as imperial governor of German Samoa. Earlier he had served as an attaché at the German Embassy in Calcutta. He had held other tropical postings as a young man and was a vocal advocate of German expansion in the tropics.[3] In 1913 no one was more enthusiastic about solar power as a tool of German expansion in Africa than Solf. Coal and petroleum, he knew, would be commandeered by the Germany navy and German industry at home. As new sources of petroleum were developed, German colonies would be expected to send reserves to the energy-starved fatherland, not use them for their own economic expansion. Solar power, it seemed to Solf, was the perfect solution for making distant German colonies less dependent on precious fuels that would only become more in demand once a conflict with Britain began.

Before he left the Continent for his return to Cairo, Shuman consulted with Solf about the relative merits of potential sites they might visit in 1914. The cloudless Kalahari Desert, bone dry and significantly above the earth's dusty, sun-dimming atmosphere, was one of the locales they should inspect closely, Solf suggested. Shuman agreed that the Kalahari was, at least in theory, an ideal location for a solar-power installation of great size, though how the energy would be used in so remote a place was unclear. As he boarded the train for Hamburg and the steamer that would take him back to Alexandria, Shuman was confident that he had locked up yet another contract with unlimited potential for Sun Power Company, accompanied by a generous advance of nearly a quarter million dollars. With British and German financial support already in his pocket, he felt sure it was only a matter of time before a major American project would follow.

---

[3]Energy issues were often at the heart of Solf's diplomacy. Serving as Germany's ambassador to Japan in the 1920s after the war, he observed that the Japanese island people had been stymied as a culture and a modern nation by indigenous energy shortages.

In spite of all this success, or perhaps in part because of it, Shuman was becoming homesick, and he thought it was an opportune moment to consider taking his case back to the United States, a time to take new bearings on the future, to think about what his next move should be, and perhaps to retest East Coast financial sources for projects closer to home—in Arizona, Colorado, Nevada, California, or other part of the great American West. But he still had work to do in Egypt.

Solar Power in the Sudan

Back in Cairo, Shuman quickly became involved once more in operations of the Maadi plant and efforts to fine-tune the energy output of the absorbers and the engine. He also continued to build on his Maadi success to pursue the new opportunities in British North Africa Kitchener had promised, especially in the Sudan. The Maadi plant, relatively small compared to Shuman's original ambition for it and still plagued by the kinks and hiccups any prototype is prey to, was hardly the last word in solar-irrigation plants. Shuman wanted Kitchener to know that he was capable of much more—bigger, better, more powerful installations. The Sudan—infinitely more remote than Lower Egypt, with dry air, unlimited sunlight, and absolutely no indigenous energy sources within hundreds of miles—might provide the circumstances for a truly colossal solar irrigation experiment.

The Sudan was almost a continent unto itself. It covered a million square miles (almost the size of Alaska, Texas, and California combined), by far the largest single territory in Africa. Until the last couple of decades of the nineteenth century, it was largely unexplored. It was remote, even from Egypt—Khartoum was more distant from Cairo than Alexandria was from Rome. The Sudan stretched from the 22nd parallel of latitude at Wadi Halfa in the Nubian Desert all the way to the equator. Its borders, excluding that with Egypt and the coastline along the Red Sea, extended for more than 4000 miles. To the east this frontier was well defined by the Red Sea itself and by Ethiopia, but to the south and west the vast territory became lost in swamps and deserts with the virtually independent sultanate of Darfur in the northwest and unknown and unvisited tribes in the southwest.

Until the rebellion of the Mahdi[4] and the fall of Khartoum in 1885, the Sudan had been a province of Egypt with a population of about 8 million. By 1899 when Sir Reginald Wingate arrived as governor-general of the territory, that population had fallen to fewer than 2 million. British retaliation for the death of the British commander of Khartoum, much of it inflicted on the Sudanese population by Kitchener and his troops, had been swift and brutal; the country, such as it was, had been destroyed. There remained "a few moldering villages and some scanty cultivation," in the words of Wingate's son, who accompanied his father to his post. "Of government there was none, and of indigenous personnel to form a government, or even to staff the lowest post in any administration, there was also none."

Khartoum itself was in ruins. Sand, bush, and jungle had overwhelmed its surroundings. The huge slave city of Omdurman, the site of Kitchener's decisive victory and once a town with a permanent population of 20,000 inhabitants, mainly Arab traders and a transient population of up to a quarter of a million slaves penned in corrals and warehouses awaiting transshipment by caravan to waiting dhows on the Red Sea, had been reduced to a "sprawling mass of hovels stinking to high heaven." Sudan had to be "recreated anew." For this task two men, first Kitchener and then Reginald Wingate, each in his role as governor-general, had been given supreme civil and military power by the British government.

Wingate, like Kitchener who had preceded him in Khartoum, was not a product of the elite British public school system; the playing fields of Eton had played no part in forming the character of

[4]The term *Mahdi* in Arabic simply means "he who is divinely guided" and in Sunni Islam signifies one who is sent by God to restore faith. Sunni Muslims believe that the Mahdi will appear at the end of time to restore justice on earth and establish universal Islam, a concept not unlike the Judeo-Christian notion of a messiah. Among Shiites, the concept of the Mahdi has centered more on individual imams who have claimed this title, and throughout Islamic history many reformers proclaiming themselves the Mahdi have arisen. The most famous was Muhammad Ahmad, a Sudanese leader who said he was the Mahdi in 1881. He died shortly after successfully attacking Khartoum and killing General "Chinese" Gordon, the British commander there. In British accounts of Sudan's history, references to "the Mahdi" always mean the person of Muhammad Ahmad.

either man. Both had risen in the British army in part because of their willingness and ability to take on onerous, dirty jobs that better-connected Englishmen could avoid. Commissioned in the British artillery in 1880, Wingate was assigned to the Egyptian army in 1883. Six years later he became director of Egyptian military intelligence, a job one of his peers described as "imprisonment without term in a dark sewer." He fought in several battles against adherents of the Mahdi, but he made his name and his career on November 24, 1899, when he defeated and killed Khalifa Allah bin Muhammad, the military and religious successor to the Mahdi in Sudan, greatly impressing Kitchener and the British administration in Cairo, who had been plagued by Khalifa's military incursions for more than a decade. The next month Wingate was appointed governor-general of the Sudan and simultaneously sirdar of the Egyptian army, the post previously held by the man who had by now become his principal friend at court, Kitchener of Khartoum.

Sir Francis Reginald Wingate had effectively become the absolute ruler of the Sudan, and except for financial controls exercised from Cairo, he was left pretty much alone. The power relationship between Cairo and Khartoum had been established years earlier when Sir Evelyn Baring was British agent in Cairo and Kitchener was in Sudan. "Generally, I want to control the big questions," Sir Evelyn had instructed Kitchener, "but leave all the detail and execution to be managed locally. By the 'big,' I of course include," he added significantly, "all such measures, for instance, as involve any serious interference with the water-supply of the Nile, or any large concession to Europeans or others."

Kitchener, unlike Baring, would have much preferred to micromanage Sudan from Cairo, but by then the protocols were well established and it would have been hard for him to reverse Baring's policy, which he had created to a large extent to give himself a free hand. And Kitchener's own superiors back home, he knew, really had little time for Sudan. The Sudan and the scandal caused by General Gordon's death (he had been more or less abandoned to his fate by a penny-pinching government) were still regarded as a shameful memory in London, an inexcusable military defeat at the hands of men the British considered savages. The principal aim of Sudan's governor was to avoid making headlines in London and to keep a lid on native uprisings while doing what he could to improve the lot of the Sudanese and, if possible, make some money for the crown.

It was Wingate, over a period of nearly a decade and a half by the time of Shuman's Egyptian adventure, who had established peace and order across the territory and who had manufactured a civil service administration almost from scratch. Unlike the men from whom he took orders (possibly excepting Kitchener), Wingate had ambitious plans for the Sudan, which he thought could become a valuable addition to the empire. Kitchener encouraged him in this view, especially since he thought that a combined Egypt and Sudan might be considered important enough to be made a British viceroyalty, headed of course by himself.

Irrigating the Sudan to Grow Cotton

To bring Sudan back from the edge of chaos, Wingate had to find a way to create economic activity. This he proposed to do through agriculture. In the second decade of his governorship, after he had quelled the continuing political unrest in the wake of sporadic Islamic insurrections, he developed a great plan for the irrigation of the area known as the Gezira—the triangular expanse of land south of Khartoum bounded by the Blue and White Nile—to grow cotton. The scheme called for crisscrossing an area of nearly 10,000 square miles with interlinked canals. Because of the V formed by the two rivers, no point of land in this huge area would be more than 30 miles from the banks of one of the branches of the Nile. Canals could reduce the distance to water to no more than a mile or two.

Gravity alone, however, would not suffice to move the water; pumps would be needed. By 1911, two years before Shuman visited him, Wingate had ordered exhaustive experiments to establish whether long-staple cotton, which grew so well up north in the Nile Delta, could be grown in the Gezira as an economic crop. These experiments had demonstrated by 1913 that conditions for cotton in Sudan were almost perfect—if sufficient water could be provided.

Cotton plants had "an almost unquenchable thirst," Wingate told Shuman. Cotton required an ample supply of water during the summer, which the traditional systems of irrigation in Sudan (just as in Egypt) could provide only through the lifting of water by hand- or animal-driven pumps. If agriculture in Sudan was to flourish, it would be dependent upon a bounteous water supply, and practically every thimbleful of water would have to be brought to the land from the Nile by some artificial means. Further, for cotton, sufficient water had to be made available at all times of the year in the exact amounts required. A day or two of drought could be fatal. At the end of the growing season when the plants were large and the fields were densely covered with foliage, 50 tons of water a day per acre could easily be soaked up by the cotton.

Not surprisingly, the success of cotton in North Africa was a fairly recent phenomenon. At the beginning of the nineteenth century, commercial cotton in Egypt was almost unknown, though some forms of cotton agriculture had been recorded in Egypt since Ptolemaic times. Yet by 1912 production of cotton linked the agricultural fortunes of Egypt to those of every Western nation. It was said by British journalists that when European businessmen thought of Egypt, they thought of only two things—the Suez Canal and the special kind of cotton that was grown there, said to be the best in the world.

Wingate was as aware as Kitchener that cotton had become the chief source of Egypt's income. He hoped to piggyback cotton production in Sudan on that success, taking advantage of the growing demand Egypt had created but could not alone fulfill. The market was large. Egypt's export of raw cotton was worth between 20 and 30 million pounds sterling a year, much of it shipped to Great Britain. England held the primary interest in Egypt's cotton crop, spinning nearly half of the entire Egyptian yield in Lancashire alone. Though the British did not advertise the fact (preferring to focus on their machines and high-technology products), manufactured cotton goods constituted in 1912 the single largest export product from the whole of the British Isles. The Egyptian supply was used mainly in the top end of the market by "fine" spinners whose products were the most prized and the most costly. Though Egyptian cotton constituted only 6 to 7 percent of the world's cotton crop (more than 50 percent came from the American South), it commanded the highest prices because it could be spun into the most luxurious grades of finished goods.

Demand for cotton in the late industrial revolution was such that most of the world had been surveyed for cotton-growing potential, but no region of the world could quite compare with Upper and Lower Egypt in providing ideal conditions. The kind of cotton grown in the Delta in ancient times was known as "tree cotton," an inferior kind of cotton not unlike that produced in India. In 1820 another variety of cotton of better quality was transplanted to Egypt. Called Jumel cotton, it too was a perennial like tree cotton. Before the Civil War, cotton plants were imported from Georgia and Alabama in the United States and from these a special Egyptian hybrid soon supplanted earlier varieties. Growth was explosive, in part because of shortages created by the Civil War itself, which had strapped British mills for raw material.

Egyptian production was explosive in the nineteenth century, growing from about 100,000 pounds of cotton lint in 1821 to nearly 23 million pounds by 1824. By 1852 production had grown to 67 million pounds, and production hovered at about that level until the outbreak of the American Civil War in 1861. In that year Egypt produced 60 million pounds of cotton. Spurred by American shortages, by 1865, just 4 years later, Egyptian production had nearly quadrupled to 215 million pounds. The Egyptian cotton boom was, in a sense, a gift from the United States. Cotton prices had also soared, from 5 to 10 British pence per pound before the war up to between 22 and 32 pence a pound in 1865—a 500 percent rise in price in some instances.[5] The boom in cotton production and the price bubble both came to an end with Lee's surrender, but Egypt by then had captured a major slice of the high-end market, with production remaining at about 200 million pounds of cotton lint per year steadily thereafter.

The growth of Egyptian cotton production by the end of the American Civil War was constrained only by the natural limits

---

[5]British currency in the nineteenth century consisted of pounds, shillings, and pence. There were 20 shillings to the pound, and 12 pence to the shilling. While a pound sterling might have secured nearly 50 pounds of cotton lint before the Civil War, it bought the British mill owner less than 8 pounds of the raw material during the war in the extreme case.

imposed by the Nile's water supply; further expansion of the crop had to wait for improvements in irrigation. These did not come until the British occupation, after which British demand alone forced production to grow. By 1900, the crop exceeded 600 million pounds and by Shuman's time a crop of 1 billion pounds seemed feasible if all the Nile's resources could be efficiently deployed, though the largest annual crop produced before Shuman proposed his solar water projects was only 770 million pounds.

By the time of Kitchener's arrival in Egypt as consul general, the Egyptians had developed problems in cotton agriculture that Wingate quietly hoped to exploit. Efforts to increase Egyptian production by extending the summer crop had been disappointing, largely because average per-acre yields had begun somewhat mysteriously to fall. The cause of this yield drop was controversial: Some British agronomists traced the origin to lowered soil fertility caused by an oversupply of water without sufficient drainage, but Egyptian farmers and even British colonial officials—in a land where water was considered a gift of God—refused to accept an explanation that centered on a surfeit of water. By 1910, the trend in fertility loss in the soil seemed to be turning around on its own; the average yield per acre was rising from a low of 450 pounds per acre to its old level of about 550 pounds per acre. A 1-billion-pound annual Egyptian crop was back in the sights of British planners in Cairo.

Notwithstanding these positive developments, there was a definite upper limit, in Wingate's long-term view of the situation, to the potential size of any further growth in Egyptian cotton production, at which point Egypt's market share would necessarily begin to shrink if world demand continued to grow as expected. It was into this widening breach that he proposed to launch his program of Sudanese cotton production without creating too much of a furor in Cairo and London. He would be filling demand that the Egyptians could not hope to meet on their own—and therefore could not object to Sudan filling.

Wingate hoped that Sudan might provide an even better growing environment for cotton than Egypt, if he could solve the water supply problem. There was an ample supply of hand labor in Sudan (as in Egypt). The best cotton could not be grown in fields plowed by horses or tractors; handheld hoes were essential if the plants were to be kept optimally close to one another. The harvest of cotton had to be picked from the open fruits by hand—a labor-intensive process that could cost the farmer half the value of his crop. The small-holding Egyptian fellahin, incredibly industrious and with large families, provided this labor force in Egypt. Wingate was confident the Sudanese would be as successful as the Egyptians, if water could be supplied at low cost.

There was also the question of temperature and climate. The delicate hairs of the finest cotton seemed to achieve the very finest quality when conditions were such that growth proceeded rapidly at night, followed by a kind of "hardening off" in the harsh heat of the day; the British referred to this as "checking." Under Egyptian conditions, the checking of growth by dry hot air and a pitiless sun was an important condition in the fruition of the perfect crop. Were it not for these tough conditions, the cotton plants would sometimes reach 20 feet in height instead of the optimal 6 feet, and much of the plant's energy would have been misdirected to the stalk and leaves instead of to growth of cotton fibers. On the other hand, if daytime checking was carried too far, even for a few days, the cotton plants would be permanently injured. Conditions in Upper Egypt as well as the Sudan struck exactly the right balance between hot, dry days and cool, moist nights to allow cotton to flourish.

British "Hydropolitics" Closes the Purse

Just before he met Shuman, Wingate ordered Sir William Garstin to Sudan to head his irrigation survey and take charge of the vast Gezira project. Garstin brought with him nearly 20 years of experience on the Nile. Before the turn of the century, he had traveled throughout northern Sudan and steamed up the White Nile to see for himself what parts were navigable. Except for short stretches of open water, he had found the river completely clogged by vegetation. Garstin reported to Cairo that there could be no question of effective occupation of southern Sudan until the Nile and its tributaries were cleared. While serving as inspector general of irrigation at the Egyptian Ministry of Public Works, Garstin proposed the 150-mile "Garstin Cut" to channel the White Nile through the Sudd through a human-made canal, a proposal now being revived by Wingate.

In the intervening years, the northern portion of the canal had actually been built and it stood now half-finished, 90 miles long, filled with stagnant water and breeding billions of mosquitoes. British engineers were enthusiastic designers and advocates of improvement projects in Sudan, but imperial "hydropolitics," which put strict limits on the amount of British treasure that could be committed to an area as remote as the Sudan, blocked development of most of these initiatives.[6]

Though he had been seconded to Wingate in Sudan, Garstin maintained close contact with his colleagues in Cairo, who urged caution in implementing the survey. Proposals to tamper with the Nile's flow, however well intentioned, could not fail to create jitters in the British Agency and probably also in London. The summer Blue Nile flood was the source of all of Egypt's prosperity, and it was imperative that any water drawn off for Sudan irrigation not damage Egypt's vested interests.

Garstin eventually came to champion a plan to dam the Blue Nile at Semnar (160 miles south of Khartoum) and build canals with reinforced high banks to carry the water northward for irrigation. But Garstin, for all his expertise, had little influence with British politicians who would have to approve the expense. It was up to Wingate to get the money for the proposal at a time when the British treasury already viewed Sudan as a boondoggle.

Wingate put together a private consortium of British farmers who had been operating a small pilot cotton scheme in the Gezira to back the dam and put up some of their own money. He secured the support of Sudanese landowners by proposing a profit-sharing agreement of 40 percent to the Sudanese tenants, 40 percent to the

[6]Earlier, in 1876, one of Garstin's predecessor members of the British Royal Engineers, General F. H. Rundall, had even proposed a high dam at Aswan, a dam that would remain a dream until the 1960s.

British government, and 20 percent for the two private investors. With these agreements in his pocket, Wingate hoped to persuade the British government to underwrite a bond issue in the London debt market to raise 3 million pounds to actually build the dam. He went to London and made his case. In spite of the support of the British Cotton Growing Association, whose chairman Wingate had invited to Gezira and who gave the project a glowing assessment back in London, and notwithstanding Kitchener's personal endorsement, it seemed that the new proposal would share the fate of all previous proposals that entailed any major British financial commitment.

When the project finally seemed doomed, Wingate happened to be invited to visit the royal family at Balmoral for a short holiday. He mentioned the dam project to the king, who told him that Lloyd George, then Asquith's chancellor of the exchequer, was arriving the next day. In a last-ditch effort, Wingate tried to get 10 minutes alone with Lloyd George every day for 5 days running but was rebuffed by aides who had been warned of Wingate's proposal and who knew that Lloyd George had a soft spot for grandiose engineering schemes. On the last day, Wingate corralled Lloyd George alone over breakfast—both men were early risers and their assistants were sleeping—and got the finance minister's personal commitment for the bond issue. The news electrified Cairo and Khartoum—3 million pounds for Nile irrigation was an unheard-of amount of money to be committed to such a remote colony, no matter how impressive the plan to spend it.

In the meantime, the world edged toward war. By the end of April 1914, Wingate was back in the Sudan, and King George, accompanied by Foreign Secretary Sir Edward Grey, was in Paris to argue with the French about an Anglo-Russian naval convention that the British were loathe to support, despite pressure from France to undertake it. Discussions dragged on until May. By early June, Russia's Czar Nicholas was in Bucharest to conclude an agreement that called for the Russians and Romanians to cooperate in the event of the closure of the Bosporus Strait in a Turkish-Greek war, though he was unable to get the Romanians to commit themselves to intervene in the event of an Austrian attack upon Serbia. And in Sarajevo, Austria's designs on Serbia were very much on the mind of an obscure young Bosnian revolutionary, Gavrilo Princip, a member of the Union or Death terrorist group, who was building a bomb with which he hoped to assassinate Archduke Francis Ferdinand, if he ever got the chance.

By the time he returned to Cairo to make preparations for home leave in the United States, Frank Shuman had a firm commitment from Wingate and Kitchener for 30,000 acres of prime Sudanese agricultural land to be dedicated to solar-powered irrigation and farming, the largest solar demonstration project ever contemplated. But would the British government, so sensitive to anything that might disrupt the delicate balance of Egyptian politics and economics, go along? Shuman learned of Wingate's triumphal breakfast with Lloyd George while he was sailing on the SS *Mauretania* homeward bound, heading west for the United States from England. He was thrilled. Now he had more reason than ever to regroup and to celebrate his good fortune back home in Philadelphia. He missed his inventor's compound in Tacony. He had been on the road more than 2 years.

Though war might be on the horizon, Frank Shuman had accomplished more than he had ever dreamed possible when he set out for England in the Cunard flagship in the fall of 1911. By hitching his star to massive government-sponsored irrigation projects in Africa, both British and German, he had maneuvered solar-power technology into a position where it seemed to be coming into its own at last, and on a colossal scale. Nothing, it seemed, could stop him now.

Solar-power development in the United States would not follow such a grand path. While Shuman had occupied himself overseas, other solar-power inventors had been active in America. He would find the competition at home tougher than when he left.

8
California Light and Solar Power

Frank Shuman had considered any number of sunny U.S. locations for large solar-power demonstration projects—Florida, Arizona, and California, among others. It was the search for money, mainly, that drove him to London and Egypt. In the wake of John Ericsson's well-publicized experiments, dozens of American engineers and just plain tinkerers had taken up the call of solar-generated power in the United States. Many were active in the two decades before and after the turn of the century, including at least one interloper from Great Britain. This was Aubrey Eneas, who set his sights on California's clear skies and bright sunshine.

In 1903, Eneas found the perfect Southern California location for his Solar Motor Company: a five-story, red brick and sandstone building with an Italian Renaissance façade, at the corner of Broadway and Third Street in downtown Los Angeles. He chose this building because of its great light-filled court rising 50 feet above the Belgian marble stairs, ornamental cast-iron railings, Mexican tiles, polished wood, and the black metal fretwork of the open cage elevators—the most recent and by far the sleekest model of Elisha Graves Otis's landmark American invention. And an enormous skylight (made possible by wire glass, Frank Shuman's fortune-making brainchild) filled most of the roof of the building.

Within the airy enclosure below, ethereal light gave the glazed brick walls a weightless quality as though the whole building might simply float away on California sunbeams. It was a building turned inward on itself, with the exterior light miraculously imprisoned within. Its interior was dazzling. The court was flooded with light, all of it natural. In their freestanding shafts surrounded by swirling dust motes and intricate metal grillwork, two elevator cages rose toward the roof like giant beanpole trellises. Geometric, patterned staircases shaped from French wrought iron and Belgian marble balanced the elevator cages at either end of the atrium.

The nineteenth century had just given way to the twentieth, and this was the Bradbury Building, built in 1893 by a California mining tycoon turned real estate entrepreneur, Louis L. Bradbury, who erected it just a few blocks from his baronial home on Bunker Hill as a memorial to himself. It would cost him three times what he had budgeted for it, and he would not live to see it completed. The Bradbury Building, along with fashionable Bunker Hill and much of the business district, had all the latest gadgets and comforts modern engineers could offer in the last decade of the nineteenth century. Buildings were linked by telephone, and electrical power was available to most. Piped, running water and hot water in the washrooms (often produced by solar panels on the roof) were common conveniences.

The Bradbury Building, built for the fantastic sum of a half million dollars, was at the turn of the century the poshest business address anywhere along the Pacific coast, the ideal venue for any ambitious entrepreneur who wanted to make a good impression on financial backers and clients alike. But it was the magical, otherworldly light that bloomed within the interior of the building, more than its prestige as a business address, that inspired solar inventor and business promoter Aubrey G. Eneas to locate his Solar Motor Company offices there.

A Six-Story Solar Machine

Aubrey Eneas was temporarily flush with cash, an unaccustomed condition for the British-born engineer from Massachusetts, and he could afford the new digs, at least for a while. Solar power had already had a receptive audience among the farseeing Californians; by the turn of the century, 30 percent of Pasadena's larger homes boasted solar water heaters.[1] Eneas brought to California the financial backing of a number of unnamed "Boston capitalists." More important, he had just sold, for the princely sum of $2160 (a comfortable annual salary in those days), a complete solar machine including reflector, boiler, steam engine, and pump to Dr. Alexander J. Chandler, a Canadian who had amassed more than 18,000 acres of Arizona ranch land outside Mesa, 35 miles southwest of Phoenix, in the Great Western Desert.

The device Eneas sold Chandler was a six-story monster weighing 8300 pounds. It used a truncated cone-shaped reflector (somewhat along the lines of the device Mouchot had exhibited in Paris) to collect solar radiation from about 700 square feet of surface area. The cone was nearly 34 feet in diameter at the top, narrowing to 15 feet at the bottom. It held 1788 flat mirrors packed along the curved surface like the tiles of a mosaic to direct the sun's rays to the central boiler. The boiler was made of two concentric steel tubes. These in turn were encased in two glass tubes with an insulating air space between them. A second air space between the inner glass tube and the outer steel tube provided yet more insulation. The water circulated up between the inner and outer steel tubes, where it was warmed by the concentrated solar rays, and then down the inner tube.

The boiler was placed at the axis of the cone. It was a large device, some 13 feet 6 inches in length and a foot in diameter, shaped like a torpedo or cigar. It had a capacity of nearly 900 pounds of water (10 cubic feet) with a steam space of about 8 cubic feet. In full sunlight the boiler quickly developed 150 pounds per square inch of steam pressure. The cone-shaped mirror concentrated sun-

[1]Solar water heaters in southern California usually held from 20 to 140 gallons and could be heated on a sunny day to between 115 and 140 degrees Fahrenheit—a temperature more than sufficient for most household uses. The simplest devices were metal tanks painted black and placed to maximize exposure to the sun. More sophisticated models resembled the solar heaters found on California rooftops today—flat hot boxes in insulated wooden frames covered with glass (much along the lines of the Shuman system) to let heat in but not out. These were capable of producing hotter water.

One of Aubrey Eneas's solar machines in Arizona, set up on the John May farm to pump water for irrigation.

light by a factor of about 14—that is, nearly 14 square feet of sunshine were concentrated on each square foot of the external surface of the boiler.

Surviving photographs of the installation show a steel-girded contraption resembling a modern-day radio telescope. The device could develop 11 horsepower and pump 1400 gallons a minute of Salt River water onto Chandler's parched fields. To keep the cone trained on the sun, the intricate assembly of glass and iron which made up the conical mirror and boiler was harnessed to a clockwork mechanism that kept it all moving around a pivot to track the sun as it moved across the sky.

Chandler, like Eneas, was a man of big dreams and small cautions. His dreams were large enough to include the possibility of solar power being used to irrigate immense tracts of Arizona desert. An unlikely immigrant to the Southwest, Chandler had been born in Quebec in 1859, the son of a Baptist minister. As a youth, he had fantasized about leaving the cold winters of his childhood to travel to the American West, where he dreamed of becoming a cattle rancher and land baron. He graduated from the Montreal Veterinary

College at McGill University in 1882 and started a practice in Detroit, Michigan, abandoning Canada for good. He became an American citizen and started making money and business connections in his new homeland. His real ambition lay in the American Southwest, and he kept his eyes trained there for any opportunity that might present itself.

Five years later he found the opening he had been waiting for. Arizona's newly created Sanitary Livestock Board hired him as the territory's first veterinary surgeon. He arrived in Prescott, the territorial capital, in August 1887, at the height of a severe drought. Discouraged by the lack of water and appalled by the toll it was taking on local livestock, he resigned his position after a month on the job, convinced that the cattle industry had no future in that part of the world. But by then he had fallen in love with the warm climate, and he refused to fold his tent and go back East. Instead he took the Black Canyon stagecoach to Phoenix to catch a train to California, where he thought he might find better outlets for his veterinary skills. And then, the night before he was due to leave Arizona, it began to rain. The deluge lasted three weeks.

In the ensuing days Chandler, like so many before him, was astonished and enchanted at the changes wrought in the desert by the now plentiful water. He rescinded his resignation and canceled the trip to California. Soon he was the owner of two Salt River Valley ranches of 160 acres each south of town, the core of what would later become the huge Chandler Ranch. The Chandler properties were primitive places. One adjoined a village (later the town of Chandler, named for him) consisting of three wooden shacks: the townsite office, a dining hall, and a grocery store. And yet where others saw only boardwalks and tumbleweed, Chandler saw a burgeoning city surrounded by fertile farmland.

Irrigating the Arizona Desert

The key to the transformation, of course, was water. In the next few years, Chandler worked on ambitious plans to transform the arid desert. He laid out, if only on paper, landscaped parks and cotton fields, and he foresaw a growing population attracted by the new wealth he would create. The land was there and available, and with water the desert would certainly bloom.

The difficulty in getting plentiful water for his Mesa ranch, even at those times of the year when the Salt River was in spate, led Chandler to become interested in irrigation engineering and to look into building dams and canals around his property. Though the area already had a network of primitive canals, these relied mainly on brush dams laid out by amateurs using mule teams, slip scrapers, and wagons along with the occasional charge of dynamite to move a boulder too heavy to shift with a crowbar.

Chandler's reading about massive engineering projects like the Suez and the Panama Canals (the latter was still being built) taught him what modern methods could accomplish in transforming the surface of the earth to make water flow where people wanted it to flow. All that was needed was sufficient capital, mammoth dredges, digging machines, electric pumps, and trained workers. Using his new knowlege, Chandler designed an irrigation system that dwarfed in scope and efficiency the one created haphazardly by the early settlers of Mesa, Lehi, and Tempe.

Previous settlers in greater Phoenix had formed the Mesa Canal to tap the Salt River, but it was an inefficient, often stagnant waterway. Chandler formed a consortium of local ranchers to create a new Consolidated Canal Company in 1892. The consortium promptly took over the Mesa Canal and began excavating the Crosscut Canal, 2 miles long, and its extension, the Tempe Canal, to tap into the flow of the Salt River. Chandler's plan was to divert a portion of Salt River water at a permanent masonry diversion dam erected at the confluence of the canal system and the river. The Tempe Canal water would then be carried by the Crosscut to a point on a bluff on his property where it would drop off the mesa, turning electric generators in a power plant he proposed to build there, before it was diverted into irrigation ditches from which it would soak the parched earth for miles around.

Because of the constant need to rebuild the crude brush dams after every minor flood in the Salt River, the owners of the Mesa

Canal Company were soon convinced of the wisdom of turning over their canal system to Chandler, who guaranteed them that his permanent diversion dam would ensure the constant water supply the area needed to flourish.

Not everyone liked the arrangement. An early settler who had come from Utah with a Mormon group in 1878 complained that "our people have made a grave mistake in yielding up their water system," predicting that area residents (other than Chandler) would be reduced to "small farms, and a rental on their water." Other residents were concerned that so much of the capital—more than $1 million—had come from Detroit financiers, absentee investors who were friends of Chandler from his earlier stay in Michigan. Water being scarce in Arizona, the simple reality was that Chandler's water gain represented someone else's loss. He was soon embroiled in a series of lawsuits contesting water rights. They would plague him for decades.

Undeterred by what he considered a nuisance, Chandler set his sights on enlarging his empire by acquiring the thousands of acres of unoccupied desert south of Mesa nominally owned by the federal government. He knew he would have to act quickly, because if he succeeded in providing ample water for his own properties, as he surely expected to do, the market value of the surrounding desert would soar.

Chandler developed an ingenious land scheme that involved inducing others to acquire the unoccupied federal lands as "dummy entrants" who would later convey the acreage to him. The strategy came to light in 1912 (the year Arizona achieved statehood) when Chandler was hauled before a congressional committee in Washington to testify about his land acquisitions. Before the hearings were over, Congress determined that he had defrauded the federal government of more than 18,000 acres, but by then it was too late to prosecute him or the people he had induced to help him in the swindle, or even to repossess the acreage.

It was a simple, effective scheme, and it made Chandler one of the biggest landowners in the nation. With the help of a Phoenix lawyer, Chandler interpreted a provision in the Desert Land Act of 1877 to mean that investors could issue mortgages on federal land in

Arizona. Courts later ruled there was no such provision, but at the time it was good enough to provide the basis for his plan to accumulate land. Chandler enlisted local people, many of them former employees, to help him.

He laid out an attractive proposition: Chandler would pay all costs of making application for federal land, for clearing the land and putting in irrigation ditches, and he would provide irrigation water through the canals he owned. In addition, each participant who made an entry for 640 acres of land from the government (1 square mile—the maximum allowed per person) would be guaranteed 40 acres free and clear, along with water rights to irrigate it. In exchange, the dummy entrant agreed to mortgage the land to Chandler at the exceptionally high rate of $25 per acre, or $15,000 for the 600 remaining acres. In other words, if Chandler held up his end of the bargain and provided irrigation water for the land, his proxies could pay Chandler $25 an acre to keep it. If they didn't have the $15,000 for 600 acres or didn't want the land, they agreed to convey it permanently to Chandler upon receiving title for the acreage from the federal government.

In an era when $100 was good wages for a month of blue-collar work, it was ludicrous to expect carpenters, gardeners, housewives, and the other citizens of modest means whom Chandler had enlisted to accumulate in 3 years (the period within which, by contract, water had to be brought on the land) $15,000 to redeem the mortgages. Chandler, of course, wound up with all the acreage, which is just what he'd planned all along.

Chandler, meanwhile, continued to improve the network of canals that was to irrigate the nearly 30 square miles of land he would soon own. With backing from his Detroit investors, he dredged and widened the Mesa Canal. He installed a new canal head, which had to be blasted through 1300 feet of solid rock. The head gate, about 150 feet from the river, was set in huge concrete blocks. The *Tempe Daily News* reported that a dredge weighing 300 tons, said to be the world's largest, was used to excavate the canal from the river to the division gates. Just as Chandler had dreamed, a world-scale water development project was transforming the face of Arizona around Tempe and Mesa.

Though the canals moved water efficiently across the countryside, from the confines of the Salt River into the dry areas beyond reach of the river, that solved only half the problem. The Salt River was surrounded by a plateau. How was Chandler to lift the water out of the canals and onto the plateau where it could do the most good? It would have to be pumped.

In turn-of-the-century Arizona, where wood was as scarce as snow and where coal had to be brought in by rail all the way from Pennsylvania, any alternative to burning fuel was worthy of the serious attention of a rancher planning irrigation on a large scale. Conditions were right for cost-effective solar irrigation, and Aubrey Eneas claimed that he had just the machines to do the job.

The installation of the Eneas machinery was made in the summer of 1903 by an engineer sent over specially from the Bethlehem Steel Company. In a 1974 interview in the *Tempe Daily News*, nonagenarian Carl Spain, a longtime local resident, recalled the installation of the cumbersome solar device:

> *About the time I was in the first grade, one Sunday my father and I walked out the Maricopa road about a mile out to where they were building the solar affair. The giant dish took the sun's rays and reflected them on a boiler in the middle of it. I remember standing under that affair and looking up through the reflector that had little mirrors all around and seeing the boiler and the pipe that came down to the steam engine. That solar furnace actually worked, drenching the irrigation ditches on the parched desert land.*

Fed by the scorching Arizona sun, the Chandler machine churned out record levels of steam and power, far beyond the levels produced in earlier tests. Thousands of gallons of fresh water inundated the parched fields of Chandler's ranch for the first time. Both inventor and owner of the machine were delighted.

A New Solar Design

Eneas's success had not come easily. His California and Arizona operations were the culmination of a decade of less profitable experimentation. "One of his first productions," the editors of *Scientific American* wrote at the time, "was a silver reflector that cost many thousands of dollars but was abandoned. The next was modelled after the Ericsson machine of 1884, but it was a failure. A third was [also a failure]. A fourth try was made, this time at Denver, which was fairly successful, doing one half the work since performed by the Pasadena model."

Eneas began his solar work in Boston in 1892, basing his early-model solar collectors on the work of John Ericsson, though Eneas's machines were built on a larger scale. By 1899 he had patented several solar motors he hoped to use on the developing American frontier. In his patent application, he described his goal: "My invention is an engine or solar generator which, while adapted for generating power to be used for any purpose, is especially intended for use in connection with irrigation of the arid plains of the west."

By 1900, Eneas had organized his first Solar Motor Company, a predecessor of the firm he would later take to Los Angeles. Like so many of the inventors whose lead he followed, Eneas wanted to make superheated, high-pressure steam akin to that produced by a conventional boiler. Eneas understood that the higher the temperature of the steam delivered by the boiler, the greater the efficiency that could be coaxed from a steam engine. Eneas wanted a machine that could produce very high temperatures, for "it has been repeatedly demonstrated in steam-engineering practice that to generate steam in quantity, the temperature surrounding the steam-generating apparatus must be in excess of 1000 degrees Fahrenheit."

No Ericsson boiler was capable of that degree of heat concentration, and so he soon abandoned Ericsson's parabolic trough design. It simply could not attain the blazing temperatures he wanted. He settled on a modified version of Mouchot's truncated-cone reflector, again using new engineering techniques and metal alloys to make the cones much larger than had been possible in Mouchot's day. This he

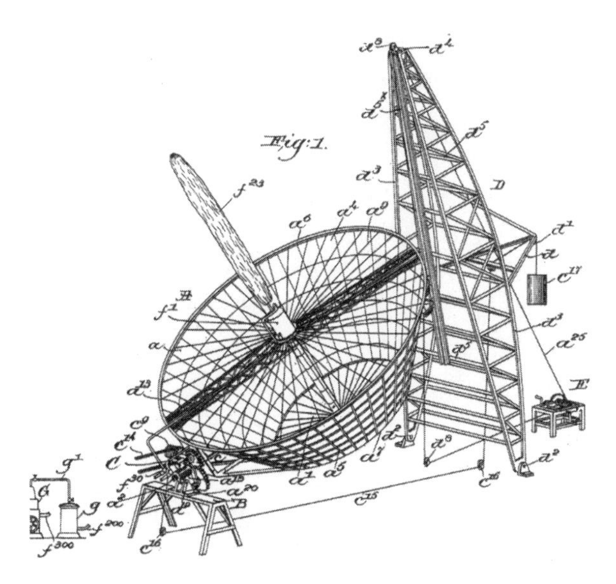

Patent drawing for Aubrey Eneas's second solar motor, 1899. To follow the sun's motion, the truncated cone mirror was moved along a track in tower scaffolding at right.

accomplished by using commercially produced, thin sheet-metal frames. These greatly increased the size of the reflector while maintaining light weight—an innovation that would prove to be the device's Achilles' heel in the windy deserts of Arizona.

Apparently, from his patent applications, Eneas was aware of the dangers posed by high winds. He explicitly gauged these risks in formulating his designs, focused mainly on the conditions he was likely to encounter in the American West, his targeted market. In a letter he wrote to his patent attorney, he explained his reasoning:

> *It is necessary in such an apparatus as this that the reflecting surface should be of considerable extent, a practical size being 32 feet in diameter, and it is obvious that the wind-pressure on such a large surface will be considerable, amounting under ordinary conditions to several tons, and also for ease of movement it is desirable that the weight of the apparatus should be as light as possible and that it should be balanced.*

In the regions of the world where this apparatus is most likely to be used, there is considerable danger from hail-storms, which of course would prove disastrous to the glass reflectors, and accordingly I have provided a canvas apron which can be extended so as to cover the reflector at a moment's notice. In certain regions of the United States where the direct rays of the sun are intensely hot—as for instance in the higher altitudes of Colorado—I find it practical to omit certain features of my apparatus and indeed it is feasible to operate the apparatus with very much less reflecting surface [than in the model he proposed for lower altitudes].

Eneas tried to incorporate enough flexibility in the design to adapt it to almost every sunny environment he might encounter west of the Rocky Mountains.

Unfortunately, although the new design resulted in higher boiler temperatures at the top of the boiler, Eneas was not satisfied with the machine's performance at the base. There the reflective area was so small (because the diameter of the cone narrowed) that a large heat differential with the top of the boiler arose. The hottest part of the boiler was found at that point corresponding to the widest part of the cone, which quite naturally was that part of the reflector that concentrated the most sunlight on the boiler. Eneas's solution was to

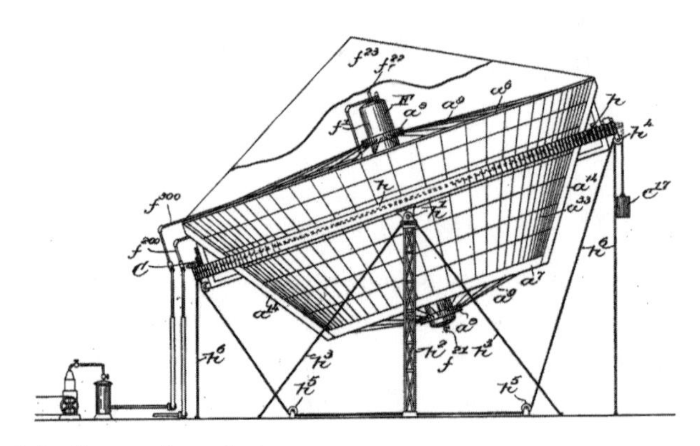

Side view of the Eneas solar collector.

cut off the bottom of the cone, eliminating that portion of the reflector that was least efficient at concentrating heat. This created a cone-shaped ring reminiscent of Ericsson's pyrometer, a shape not unlike a kitchen funnel with most of the narrow end lopped off.

Eneas's final product evolved into a truncated cone-shaped dish larger than the original design in which the sides were more upright relative to the boiler surface. This increased the amount of sunlight being gathered. It also increased the concentration ratio of the entire device. The boiler still experienced a heat differential—the hottest part was still at the top, the coolest near the bottom, as before—but it was not so steep. The average temperature generated throughout was above the 1,000 degrees Fahrenheit minimum Eneas deemed necessary to make prolific amounts of high-pressure steam. He declared he was satisfied with the design.

Field tests of these prototypes in cloudy Massachusetts—too far north for their optimal operation—produced consistently better results. Eneas soon thought it was time to test the market for his inventions in a region of the United States where they might actually be used (and purchased) by customers.

An Unlikely Solar Pioneer

Having developed a solar machine he was proud of, Eneas lacked only the means to promote it. How he did so gives rise to what is perhaps the strangest chapter in the history of solar-power technology before the age of petroleum.

On a winter's vacation in the early years of the new century in the foothills of the San Gabriel Mountains in southern California, Eneas happened to visit Edwin Cawston's ostrich farm in Pasadena. The property was a popular tourist destination for wealthy Easterners who flocked to the luxurious resort hotels in Altadena and Pasadena in December and January each year to "take the cure."

Ostriches in California were a rarity, and the Cawston farm had the ambience of an amusement park. Attendants tossed whole oranges to the giant birds, who swallowed them in a single gulp, rind and all. Crowds applauded and children grimaced as the baseball-sized lumps moved slowly down the creatures' snakelike necks. Other farmhands rode the ostriches bareback. Guests on the farm were also treated to walking tours of the property, including visits to the "feather factory." The tour ended in the giant Cawston salesroom.

Quite apart from the fame it would later earn as a result of Aubrey Eneas's contribution to it, South Pasadena's Cawston Ostrich Farm was already known across America for its impact on the fashion industry. Dyed in a variety of colors, ostrich feathers were a staple of late Victorian gowns and hats and remained so for a quarter-century, symbolizing in fashion magazines, especially in America, wealth and a kind of upper-class refinement. Imported mainly from South Africa, raw feathers commanded high prices. Cawston was the first Californian to import the birds. He helped to popularize the use of ostrich feathers, enlarging the market back East for a renewable commodity his birds were happy to supply.

Cawston hit it off immediately with his Boston visitor. He did not think solar-powered machines were at all odd; solar power was better understood in California than elsewhere in the United States. Eneas had little difficulty persuading Cawston, a visionary like himself, that solar power might well become one of the principal sources of energy to drive America's coming century. The machinery Eneas talked Cawston into letting him install on a sunny hillside of the ostrich farm was similar to the machine he would later sell to Chandler—a giant, futuristic upside-down umbrella that to modern eyes looks like nothing so much as a radio telescope. Eneas connected it to a 15-horsepower reciprocating steam engine that was soon pumping 1500 gallons of water a minute from a 16-foot borehole that Cawston had sunk below the water table. Cawston was delighted to have the extra water; Pasadena was semiarid in the days before California's first massive aqueduct was built (between 1908 and 1913) to bring the waters of the Owens River from deep within California's interior to the coastal plain and the San Gabriel foothills. Cawston used his welcome new resource to create lush gardens on parts of his property, adding yet another attraction, camellia shrubs with showy white and red blooms, for his paying guests.

Not content merely to get water from Eneas's contraption, he thought the solar engine itself might attract tourists on its own. He printed handbills to bring more guests to see his 100 ostriches and he quoted from newspapers on both coasts to support the hype he used to describe the new solar-power machine:

> *No extra charge to see The Solar Motor! The only machine of its kind in the world in daily operation! 15-horsepower engine worked by the heat of the sun!*

The solar exhibit was a sellout, although not every visitor understood the purpose of the machinery. Tour guides had to explain that there was no connection between the solar device and the ostriches, and that the high temperatures produced had nothing to do with the incubation of eggs.

Though it reduced this early display of solar technology to a sideshow, the Pasadena installation had the desired result. Word of Eneas's machines quickly reached large numbers of people who might otherwise never have heard of solar-powered pumps. Follow-up sales calls with those who expressed an interest in the technology—along with Eneas's natural gift for marketing—did the rest. Many of the tourists who visited the farm, including Dr. Alexander Chandler, left the site convinced that solar-powered pumps would become as popular in America's sunny Southwest as windmills were in Holland.

Regional California papers and America's flourishing new popular-science journals sent reporters to the farm to cover the spectacle. To Frank Millard, a reporter for the widely read *World's Work,* the potential of these new machines seemed unlimited, inspiring in him a vision of a new Eden, "where oranges may be growing, lemons yellowing, and grapes purpling, under the glare of the sun which, while it ripens the fruits it will also water and nourish them." Millard made the leap from irrigation to other commercial pursuits that demanded cheap mechanical power: "If the sun motor will pump water, it will also grind grain, saw lumber, and run electric cars," he predicted.

In spite of the circuslike atmosphere, the Pasadena machine also developed the credibility in scientific and engineering circles that had eluded Eneas previously. *Engineering News* carried an article in 1901 extolling the "perfection of the apparatus" which had been attained, it said, "not so much by any inventions as by painstaking attention to all the details of the improved application of principles which have been made use of in former, unsuccessful solar motors." The *Railway and Engineering Review* editors complimented the machine's "simple design" and seemed to like especially its portability, an achievement they found all the more remarkable for its weighing more than 4 tons: "The reflector is set in the meridian on two framed supports or towers in a manner to balance the entire frame. It rests on an equatorial mounting like a telescope, the axis being due north and south, and the machine turning east and west to follow the sun."

Scientific American published an article by Charles F. Holder, a well-known California engineer, extolling the Pasadena experiments:

> *The machine exhibited at the Ostrich Farm has attracted the attention of a vast number of people, especially as Southern California is now thronged with tourists. . . . At the time of this writer's visit to farm, the motor was the subject of no little comment. The amount of heat concentrated in the boiler by the seventeen-hundred-odd mirrors cannot be realized, as nothing can be seen but a small cloud of escaping steam; but should a man climb upon the disk and attempt to cross it, he would literally be burned to a crisp in a few seconds. Copper is melted in a short time here, and a pole of wood thrust into the magic circle flames up like a match. That the motor is a success is seen by the work that it is doing—pumping water from a well, illustrating the possibilities of cheap irrigation.*
>
> *No invention of modern times has given such an impetus in the development of arid lands as the solar motor. The development of Southern California has been seriously hindered by the lack of fuel, the country being dry and barren in localities where rich mines are known to exist. The country is cloudless for months—in every sense the land for solar motors, as water*

> *underlies the surface almost everywhere, and, when pumped up and sent out upon the soil, the region ceases to be a desert and can be made to blossom as the rose.*

Nearly 15 years later, A. S. E. Ackermann reviewed the technical performance of Eneas's Pasadena and Arizona machines and pronounced the Eneas design exceptionally efficient. "Eneas refers to his 'nine different types of large reflectors,'" Ackermann wrote, "and found that he obtained better results when he concentrated the reflected rays 'on two parts of the boiler instead of its entire length, as in the Pasadena machine.' Eneas said, 'I find 3.71 Btu per square foot per minute as the greatest amount of heat obtainable during the trial runs.' This gives a maximum efficiency of 74.5 per cent," a figure Ackermann found compared very well with earlier machines.

High Winds and Hailstorms

Based on his success in Pasadena, Eneas moved the Boston-based enterprise permanently to California in 1903, recognizing that he needed to be closer to his potential market to make sales. Not long after, he made his first deal—with Chandler. The bulky machine, transported to the burning land of Arizona, exceeded its already impressive performance in cooler Pasadena. Chandler was delighted and predicted a boom for the devices in the Western desert, where fuel was even scarcer than on the West Coast.

Unfortunately, the first machine ran into trouble after only a week of operation. After a few days, a part that was supporting the boiler at the focus of the collector broke. The lightweight rigging supporting the heavy boiler weakened during one of Arizona's frequent windstorms and collapsed, sending it tumbling into the reflector and damaging the machine beyond repair.

Though the reflector was smashed and the boiler tube was cracked, Chandler had by then watched his wonderful new machine pumping hundreds of thousands of gallons of water onto his bone-dry ranch. He was even more devotedly committed to Eneas's vision than Eneas himself was. Chandler seemed not to be discouraged by the breakdown and immediately had the machine repaired, at great expense. He was a pioneer, after all, and he was intent on transforming an entire desert.

Accustomed to setbacks himself, Eneas too appears not to have been daunted. He even sold a second machine to one of Chandler's neighbors in nearby Tempe. Six months later, in the fall of 1904, he sold a third machine to John May, a rancher in Wilcox who also paid $2500 for the complete system—solar collector, boiler, and irrigation pump, installed and ready to go.

Before its sale to May, the third machine was temporarily set up next to the railroad tracks of the Santa Fe, Prescott, and Phoenix Railroad near Tempe. It was situated, according to the *Arizona Republic*, "at the wells near the race track, one mile south and one quarter mile west of town for testing the design and making minor changes to correct deficiencies, and to publicize the capabilities of the pump before offering it for sale to the public."

The Tempe sun motor had a rather dismal inauguration on March 21, a date that was considered auspicious, being the spring equinox. The weather was cloudy that day, and continued so for the next week. A representative of the Department of Agriculture, David Griffiths, came to Tempe from Washington to see the engine in operation. The *Arizona Republic* quoted him as saying that "the principle has now been established—all that is now necessary to make it a machine of practical use is to overcome a few mechanical imperfections which will undoubtedly be taken care of very soon."

The pump worked without incident at its temporary site for more than 5 months until it was sold to John May on August 24. By September the machine had been reassembled on May's Wilcox ranch. It worked well enough, but for reasons that aren't clear—perhaps May had some premonition of what was to come—he sold the machine a few months later to a neighbor at the McCall Ranch near Cochise.

Though Eneas's patent application specified that the open-bottom design of the machine would enable the mirror collector to survive an 80-mile-per-hour wind, it was a high wind condition that seems to have brought down this machine. Sustained 80-mile-per-hour winds were a rarity even in the desert, but it was not unusual in Arizona for small "dust devils," as these miniature tornados were called, to produce even higher localized wind speeds for short periods. Dust devils in Arizona were only a few yards in diameter at their base, but the rapidly whirling wind and suspended dirt could reduce even well-built metal structures to scrap in short order. It was a small, energetic dust devil on a calm, hot day near Cochise that appears to have been the culprit that destroyed the second Eneas machine.

The Tempe machine was the largest machine Eneas ever built. It had an upper diameter of 35 feet and a lower diameter of 16 feet, slightly larger than the 1901 California model and the 1903 model built for Chandler. The price was $2500, or $3000 if installed by the manufacturer. The mirrors were of white glass (clear glass, not the cheaper, greenish glass used in window panes), silvered on the back, made especially for Eneas by Chance Brothers of London. The device was operated by winding up a large weight attached to an escapement on the highest point of the supporting tower. In the morning, the machine's attendant would point the mirror east toward the sun, wind the weight, and start the escapement functioning. The clock drive mechanism would point the mirror to the sun without further attention all day. Every few days, the attendant would shift the top pivot position to correct for the slow seasonal change of declination.

Meanwhile, the restored Chandler machine seemed to be functioning again. The installation made a lasting impression on the reporter from the *Arizona Republic* who came by to write about it:

> *The reflector somewhat resembles a huge umbrella, open and inverted at such an angle as to receive the full effect of the sun's rays on 1,788 little mirrors lining its inside surface. The boiler, which is thirteen feet and six inches long, is just where the handle of the umbrella ought to be. This boiler is the focal point where the reflection of the sun is concentrated. If you reach a long pole up to the boiler, it instantly begins to smoke and in a few seconds is aflame. From the boiler, a flexible metal pipe runs to the engine house near at hand. The reflector is thirty-three and*

a half-feet in diameter at the top and fifteen feet at the bottom. On the whole, its appearance is rather stately and graceful, and the glittering mirrors and boiler make it decidedly brilliant.

Regrettably, shortly after this newspaper account was written, this machine too was destroyed, this time by a hailstorm. Yet another weather-related catastrophic failure proved even to Chandler that the massive parabolic reflector held up by a lightweight metal frame was too susceptible to the turbulent wind conditions that plagued America's Southwestern deserts. Though he proposed a stronger, heavier, more expensive cast-iron frame to support future models, Eneas's reputation in Arizona was by now irreparably tarnished, and he was unable to make any further sales in the United States. Overseas, a machine was sold to the Egyptian government for installation in the Sudan near Khartoum, and orders were received for two more machines to be shipped to South Africa, one destined for Bloemfontein and one for Johannesburg.

In the end, the financial returns required by the "Boston capitalists" who had backed Eneas's company never materialized and they withdrew their support. Far from making money, expenditures by the company exceeded revenues at last tabulation by some $125,000, putting the Solar Motor Company deeply and permanently into the red. Almost 10 years of effort had resulted in little more than positive reviews in scientific journals and startling photographs of the machines in the newspapers of the day. The meagre sales of the 24 months he had been in business had not justified the investment required. His capital depleted, unable to coax more money from the Eastern financiers whom he had sold on the idea of solar power, Eneas had to fold the Solar Motor Company and leave his offices in the magnificent Bradbury Building. He returned to Massachusetts a broken man, never to take up solar entrepreneurship again.

The Quest for a New Design

Though the reign of the Eneas solar engines had not lasted, the idea of solar energy still gripped the imagination of the Southwest.

Unfortunately, it was a field ripe for frauds and stock manipulation. Advertisements in the newspapers of 1907 for the Solar Furnace and Power Company of Phoenix, Arizona, claimed the venture had developed a solar furnace with a hot-air engine. Potential buyers had the option of adding a conventional steam engine also. Because there is to be no record of actual working installations, it seems likely that the main product of the Solar Furnace and Power Company was stock certificates.

Other advertisements seemed to promise all but perpetual motion:

> *Any amount of water can be pumped from either deep or shallow wells: no fuel is required, and when a plant is once installed the expense is ended. Stop and try to realize what it means to be able without fire or fuel to produce heat so intense that it is hundreds of times greater than has ever before been developed by the greatest scientific men in their most powerful furnaces . . . to smelt ores and minerals right at the mines without expense of shipping often many hundreds of miles to smelter, to pump water to irrigate millions of acres of land that is now worthless, but with proper irrigation would make homes for thousands of people, to run dynamos and generate electricity to run street cars, light towns and cities, heat and light houses and cook any and all kinds of food.*
>
> *Machines can be sold for pumping and irrigation plants, various kinds of power plants, household use, heating, lighting, mining and hundreds of other purposes too numerous to mention. About a million shares left in the treasury are being offered at $2 each, and every share of this stock will in all probability sell readily for more than $2 per share. With part of this money we can buy thousands of acres of land that is now practically worthless at $2 to $5 per acre, and by installing Solar Pumping plants can sell it $10 to $100 per acre.*

Pitches like this—and the outraged complaints of bilked investors that followed—did little to help the reputation of legitimate solar-energy promoters. But the lure of limitless fuel on the West Coast was still strong, and while Eneas and the Solar Motor Company were suspending operations, another solar pioneer was just beginning his.

Henry E. Willsie, an American engineer, was as convinced as his English predecessor that America's Southwest could not flourish without cheap power for irrigation and mining. He had studied the work of Mouchot and his successors and been led to conclusions far different from Eneas's. Willsie thought that complicated solar collectors would never work. They were too expensive, and their complex construction and operation would cause them to break down in the harsh conditions in which they would necessarily have to operate in the American West. He thought that a simple solar design was key to harnessing the sun's energy economically.

Willsie began his construction of solar motors in 1904, a year before Eneas's company folded. The experiments of Mouchot and others had proved to Willsie, beyond the need to repeat them himself, that high-temperature machines relying on concentrated sunlight could not be made economic no matter how elegant and clever their design. Willsie conceded that they demonstrated sometimes spectacular technical success under *ideal* conditions, but in the real world and especially in the rugged environment west of the Mississippi, conditions were never ideal. He wanted something more sturdy, more simple—a lower-temperature collection system—but with an innovation few others had thought of. Willsie recognized that solar motors would never catch on unless some way could be found to allow them to operate around the clock. He wanted a mechanism to store the sun's heat as well as convert it into mechanical energy.

In studying the work of other inventors, Willsie was particularly impressed by the work of Charles Tellier, an aristocratic Parisian engineer. In the last decades of the nineteenth century, Tellier had devoured the scientific papers of Mouchot and Adams. About the time of their retirement in the late 1870s, he began designing his own nonconcentrating, nonreflecting solar motors. Tellier believed that Adams and Mouchot had made their machines too complicated by attempting to concentrate the sun's rays, an expensive proposition. He had a simpler alternative.

At the time he launched his solar projects, Tellier was already a famous and wealthy man. His modest niche in the history of technology was secure for his having transformed the fresh-food industry forever. Tellier had designed the first refrigerator ship, the *Frigorifique,* to carry the world's earliest mechanically refrigerated cargo: chilled meat exported from Rouen to South America. Tellier's insight was to use a low-boiling-point liquid for the refrigeration system (as modern refrigerators do), but on a massive scale, cooling the entire hold of a large ship well below freezing and making it possible to transport fresh foods vast distances without spoilage.

At first, Tellier thought he could make money selling European produce to expatriates who had resettled in South America and were nostalgic for the foods of their homelands. Though the refrigeration equipment worked perfectly, this strategy did not meet with much success from a marketing point of view. Expatriates constituted only a tiny segment of the food-consuming population, and the European merchants Tellier had partnered with soon realized the meat traffic was crossing the ocean in the wrong direction—Argentine cuts of beef, superior to the European product, have been destined for the European dinner table ever since. Once the *Frigorifique* started carrying food *to* Europe instead of *from* Europe, Tellier's fortune was assured, and several more large ships with his patented refrigeration system were built. He became a wealthy man overnight and soon wrote a book about his invention.

In 1885, Tellier installed a solar collector on the sloping porch roof of his workshop in Auteuil, the tony Paris suburb where he lived. The 10 collectors were similar to the flat-plate collectors placed atop homes today for heating water, though they were built to withstand considerable pressure. Each collector was made of two iron sheets riveted together to form a watertight seal. Instead of filling the plates with water to make steam, Tellier chose ammonia as a working fluid, mainly because of its lower boiling point. (Tellier was already familiar with ammonia as an efficient heat exchanger from his oceangoing refrigeration enterprises.) He feared that flat-plate collectors could not raise the temperature of water much above its boiling point at Paris's high latitude, even under pressure.

The collectors worked beyond expectation, generating enough pressurized ammonia gas to power a water pump he had placed in his well. It pumped at the modest rate of 300 gallons an hour during daylight, filling a reservoir on the roof. Based on this small prototype, he designed larger systems to make industrial applications possible. Encouraged by his results so far north, Tellier thought he would get much better results closer to the equator, where he thought his pumps might have the greatest commercial impact. In a letter to *La Nature*, republished in *Scientific American* in October 1885, he wrote:

> *The apparatus at Auteuil raises over 300 gallons of water per hour. In warm countries, the same apparatus would raise 792 gallons a distance of 65 feet. This apparatus differs from the numerous devices by which attempts have been made to utilize solar heat by means of the Archimedean mirror, by which only*

Large cast-iron solar hot boxes (numbered 1 through 10 in the diagram) built on the side of Charles Tellier's Paris workshop. The thick metal plates were needed because the working fluid was ammonia under pressure. Tellier hooked up the pump powered by the ammonia to lift water from his well.

> *secondary heat is obtained. It is not necessary to concentrate the heat by metallic or other mirrors: the atmospheric heat is the basis of the operation, and all roofs exposed to the sun can be used for this purpose. In this manner a valuable motive power can be obtained in warm countries without loss of room. Generating plates, such as we have described, can be applied to any roof, and if we consider that with only ten such plates 792 gallons can be raised 65 feet per hour, we can easily understand that a great elevating power can be obtained by increasing the number of plates.*

By 1889, Tellier had increased the efficiency of the collectors by insulating them. He published an essay, *The Elevation of Water with the Solar Atmosphere,* which included details of his plans to use sunlight to manufacture ice. Like Mouchot, Tellier predicted that large expanses of sunlit African plains could become industrially productive through solar power. In a second essay, *The Peaceful Conquest of West Africa by the Sun,* Tellier argued that a consistent and available supply of energy would be required before France's burgeoning land holdings in Africa could be developed.

Although the world price of coal had fallen since Mouchot's experiments, fuel continued to be a significant expense in Africa as well as a drain on hard currency because France had no coal supplies of her own. Tellier calculated that the construction costs of his low-temperature, nonconcentrating solar motor were low enough to justify its use in France's colonies, noting also that his machines were less costly than Mouchot's, with its dish-shaped reflector and complicated tracking mechanism.

Despite this potential, Tellier, a shrewder (and vastly more successful) businessman than most of his fellow solar inventors, abruptly decided to pursue his refrigeration interests instead, and to do so without the aid of solar heat. Most likely, the profits and the fame he had garnered in France from his pioneering of refrigeration—a technology whose time had clearly come—proved an irresistible incentive to make still more money. Tellier cannot have failed to notice how unsuccessful previous efforts to commercialize solar-generated steam power had been, whereas refrigeration (along with air-conditioning in theaters) was regarded almost as a kind of magic in the nineteenth century—a magic people were willing to pay for. Tellier appears not to have considered the use of solar power at sea, as the rolling motion of ships would have made solar collection complex, while high humidity often cut back significantly on the amount of solar energy that reached the surface of the ocean.

As Tellier redirected his focus, France saw the last major development of solar mechanical power on its soil until well into the twentieth century, when solar furnaces in the Pyrenees restored her preeminence. Tellier's theory of simple flat-plate collectors would have to wait for an inventor on the other side of the Atlantic to revive it.

A New American Effort

Picking up where Tellier had left off, Henry Willsie, a native of Illinois, teamed up with another American engineer, John Boyle. Sensitive to the influence bad publicity had had on Eneas—as well as the spate of frauds involving solar-power schemes arising throughout the Southwest—Willsie and Boyle were determined to keep a low profile until their successful results spoke for them, putting their integrity beyond question. As Willsie drily noted, "The building of sun motors has not been a good recommendation for engineers." Criminal frauds and legitimate failures together had given the whole field a bad name.

The partners spent a decade drawing up blueprints and conducting small-scale experiments in secret. They accumulated plans, or made their own reconstructions, of every solar motor ever built in Europe and America. Most of these, as they wrote in one of their patent applications, demonstrated that "the state of the art most developed" by their predecessors used "reflectors to concentrate the sun's rays upon some sort of boiler." This is precisely where early designs had failed, they believed, because current technology did not allow any of these designs to work commercially.

They decided eventually to concentrate on Tellier-like designs that would employ a low-temperature fluid under pressure. The mechanics of the device usually involved a simple hot box in which warmed water vaporized liquid sulfur dioxide to power a small engine. They built several prototypes, each larger than its predecessor, at their laboratory in Olney, Illinois.

By 1902, they decided to try to build a full-scale solar-power plant somewhere in the American West. The reasoning Willsie put forward, in a letter, for this decision underscored the need to find a commercially receptive environment:

> *Boyle was then in Arizona, surrounded by conditions which daily reminded one of the desirability of converting the over-abundant solar heat into much needed power. To the Southwesterner, "cheap power" brought visions of green growing things about his home to stop the burn of the desert wind, and of the working of the idle mine up the mountain side.*

In Tellier's process, which also relied on a low-boiling-point fluid (ammonia, in his case), the solar collector had to be strong enough to contain the high-pressure vapor it produced. Willsie and Boyle felt that this would make the collector too bulky, too prone to leaks, and too expensive to build. They decided to use hot water to transfer solar heat from the collector to a low-boiling-point liquid in a separate system of pipes. The heat would be transferred from the hot water under normal air pressure to the high-pressure vapor in a separate, sealed system. Their insight, potentially a big cost saver, was that solar-heated water by itself did not require the expensive, reinforced sealed collectors needed if the low-temperature fluid were to be heated directly by the sun. The hot water could be circulated in conventional pipes while the high-pressure vapor was isolated in its own smaller, closed circuit.

This innovation allowed them to use what was essentially a simple hot-box collector: a shallow, rectangular wooden box covered by two panes of glass. Three inches of water filled the box. To Willsie's delight, even in the Midwest's October cold the solar-heated water was hot enough to drive a small engine. This proved once again that a sun machine did not need a solar reflector, surprising colleagues who "were skeptical about window glass taking the place of mirrors." Two months later, they repeated these experiments in Arizona, where tests showed that the collectors were better than 50 percent efficient.

Willsie and Boyle patented the system in 1903, incorporating the Willsie Sun Power Company that year to raise capital and borrow money for expansion out West. St. Louis became home to a full-sized solar-power plant built by the Willsie Sun Power Company in the spring of 1904.

The first field of collectors looked surprisingly like Shuman's, made up of sets of large, shallow rectangular boxes. The collectors were tilted south to maximize solar-energy capture. The solar-heated water percolated through the boxes to a central tank where a separate, high-pressure circuit of liquid ammonia was boiled in a heat exchanger to power a 6-horsepower engine. The cooled ammonia vapor then condensed back into liquid and circulated into the hot-water tank for revaporization.

Meanwhile, the water in the tank separately made its way back to the hot boxes for reheating. The water and ammonia circuits were thus kept in completely separate cycles, brought together only in the copper pipes of the heat exchanger, where heat energy flowed from the water through the copper and into the liquid ammonia.

No purists, the engineers found ingenious ways to combine conventional technology with their new system. For example, the plant ran on sunless days and at night using an auxiliary boiler fueled by coal. The goal was simple—to produce power as cheaply as possible while meeting the customer's need for 24-hour operation. Like William Adams in Bombay, they regarded every ton of coal saved by solar heat as a contribution to the bottom line.

Press coverage in St. Louis and New York was extensive and positive. A number of stories proclaimed the success of this 24-hour-a-day solar-powered generator. Pleased, the entrepreneurial pair moved their laboratory. They bought land in the Mojave Desert near Needles, California—reputed to be the hottest place in the country, where the sun shone 9 days out of 10.

The Needles operation saw the solar plant built and rebuilt a number of times, interrupted by slack periods when the financing dried up, including a long stretch of enforced poverty in 1906. Needles provided a windy, dusty, harsh environment that forced the inventors to come to terms with conditions that would have destroyed more delicate solar machines. But with trial and error, the faith of the two inventors in sticking to a robust, simple system was sustained.

In one 2-year test, Willsie and Boyle found that less than 2 percent of the glass panes covering the collectors broke or became so badly discolored that they had to be replaced. Unlike the Eneas machines, the Needles plant seemed to thrive in the tough desert environment of the American Southwest. A new influx of funds in 1908 allowed the inventors to make further improvements, resulting in a fourth and final evolution of the design by the middle of that year.

The collectors at the Needles plant could raise the temperature of water to 190 degrees Fahrenheit in two stages. The south-facing hot boxes had a collecting area of more than 1000 square feet and were divided into two groups. The first group of boxes, with a single glass cover, raised the water from room temperature to 150 degrees Fahrenheit. The second group, better insulated with two glass covers (an added cost), could add an additional 30 to 40 degrees Fahrenheit to the temperature of the water. The twice-heated water then passed from the second bank of collectors to the central tank, inside of which coils of pipe containing liquid sulfur dioxide were heated as the water swirled by.

The inventors switched to sulfur dioxide from ammonia because German technicians had produced a very efficient engine that used that fluid, a machine they were eager to test in the United States. It was cheaper and more powerful than the ammonia engine and made use of lighter piping and fittings to isolate the sulfur-dioxide working fluid. The new configuration had its dangers, however. Sulfur dioxide and water, the inventors knew, react explosively to form sulfuric acid. Sulfuric acid quickly dissolves most metals. After tests, Willsie declared himself satisfied with the safety precautions the pair had taken. Further experiments with the German engine ran smoothly. Power output exceeded all expectations—a maximum of 15 horsepower.

Willsie felt he could now add the crowning element that would make his invention attractive on a commercial basis: a heat-storage system that would permit 24-hour-a-day operation, a feature earlier inventors had dismissed as impossible. To store energy for future use, Mouchot had tried breaking water into its atomic components, hydrogen and oxygen. Others had suggested that excess solar energy collected during the day could be used to lift lead weights; the force of descent at night would generate power, much like the weights on a clock. Another proposal substituted water for the weights—water would be pumped during the day up a large incline into a reservoir from which it would be released at night to power a turbine.

Aware that increasing the temperature of 1 pound of water only 1 degree Fahrenheit stores as much energy as raising a pound of lead or water 778 feet, Willsie and Boyle chose to store solar-heated water as the simplest and most effective of the methods at which they looked. It was the same principle that later guided Frank Shuman.

Solar-heated water produced during the day flowed into an insulated tank. Water needed for immediate operations went on to the boiler; the rest was held in reserve. After dark, a valve released hot water from storage and passed it through a heat exchanger so the engine continued working. Tests showed the machine worked just as planned, allowing Willsie to claim that his was "the first sun power plant ever operated at night with solar heat collected during the day."

Operating costs were lower for the solar plant than for conventional, coal-fired steam engines. According to an analysis undertaken by Willsie, running a steam engine of modest size cost $1.54 per kilowatt-hour compared to $0.45 per kilowatt-hour to operate the solar installation. Given the fuel savings, he predicted the solar plant would recover its cost in 2 years. Furthermore, he claimed that tests conducted by dispassionate outsiders apparently showed that the Needles plant worked as efficiently as any solar generator ever built. And yet, even with the simplifications introduced, the Willsie and Boyle design cost $164 per horsepower delivered at a time when most good steam engines cost between $40 to $90 per horsepower delivered. Though the fuel for the solar plant was free, the higher capital investment doomed the solar machines even in the Southwest, where coal was expensive.

At this point, the money ran out and prospects of creating a going concern, without the need for continued outside financial support, seemed remote. The two inventors must have become discouraged, for there is no record of their expanding operations. Broke and dispirited, Willsie and Boyle ended their experiments in California and went back East to conventional lives as engineers as the first decade of the new century came to an end.

The Legacy of Willsie and Boyle

Despite their business failures, Willsie and Boyle had taken a giant stride toward the commercialization of solar power by demonstrating that a solar reflector was not required to run an engine and that a simple hot-box collector could drive a low-temperature motor. That, coupled with the solar-energy storage system in tandem with a conventional engine as backup, now enabled solar-power plants to operate around the clock and throughout the year, although at costs that were not then competitive.

Despite solar power's dismal string of commercial failures, proponents continued to believe that if only they could find the right mix of solar technologies, they might yet create an economic basis for a clean, unlimited power source. The baton now passed to Frank Shuman, a man who shared that dream. It was left to Shuman to build the largest solar machine yet seen, and to try to do so on a commercial basis.

9
War and Petroleum: The End of an Era

When Frank Shuman returned to Philadelphia in the spring of 1914, he took a few months off to savor his successes in Egypt and Germany. Yet while his body rested, his inventive mind could not, and he was soon working on a project to provide U.S. military submarines with an air supply while submerged. This challenge had occupied him for years, and his solution was based on the liquifaction of air under pressure. The Navy reviewed Shuman's proposals and thanked him for his work, but the invention was never implemented in naval vessels. This reversal appears not to have detered him from undertaking other war-related research. He was soon designing a giant armored tank for the Army. His shift in focus may have resulted from private fears that the coming war would put solar steam engineering into eclipse, but if he had such concerns, he gave no hint of them to his partners.

By January 1914, A. S. E. Ackermann had accumulated enough experimental data from the Maadi plant to prepare a detailed cost-benefit analysis of the operation for the editors of *Scientific American*. The results coming back from Cairo were encouraging, and Frank Shuman was eager to publicize them. They showed that Maadi-type plants could compete on a break-even basis in any sunny location where coal cost more than a few dollars a ton. To make the technical case for the "Shuman-Boys system," as the Maadi plant was now being called in the scientific press, Ackermann reproduced portions of a technical report his consulting firm, Ackermann & Walrond, had prepared for the directors of the Sun Power Company Ltd., the previous November, which took into account the costs of construction and operation of the Maadi plant and compared them to a state-of-the-art coal-burning irrigation plant of the same size and capacity.

Good-quality British or European coal delivered in the tropics cost anywhere from $15 to $40 a ton, but the Ackermann analysis showed that the Maadi plant could compete with a coal-burning plant with coal costs as low as $2.40 a ton (a price so low it was rarely to be found except at the mine mouth). Assuming coal could be purchased at $15 a ton in Cairo (not always the case, especially in winter when high demand made it more costly), the Maadi plant could save more than $2000 a year compared to an identical plant fired by a coal-burning boiler. Because equipment for a 50-horsepower solar plant cost about $8000 delivered, whereas that for a coal-fired plant of the same size cost only $3850, the fuel savings in two years would roughly cover the extra investment required for the solar plant. Two years after that (that is, four years from startup), the fuel savings would exceed the total capital investment required for the solar installation.

Ackermann's financial analysis was also favorable even assuming that all the startup capital had been borrowed; it just took a little longer for the fuel savings to retire the debt. Ackermann calculated that for every dollar per ton added to the cost of coal above $15, the solar-powered plant would save an additional $165 per year relative to its coal-fired twin. In the out years, the solar plant would operate just for the cost of its upkeep. Shuman was confident that Ackermann's financial numbers would force even diehard skeptics to take a second look at his Egyptian experiment.

Critics of solar power pointed out that no matter how well the Maadi plant worked, it still required a larger up-front investment than a completely reliable plant fired by a conventional boiler. Who would pay $8000 for something that could be had for under $4000, future fuel savings notwithstanding? Given the choice, it required faith in the future, at a time when the whole world seemed to be going mad, for an ordinary businessman to choose the solar route.

And yet, Shuman's practical success cannot be denied. Though it was a shadow of the gargantuan 1000-horsepower installation he had originally proposed, cranked up on a hot day the Maadi steam engine produced up to 75 horsepower, enough to pump 6000 gallons of Nile riverwater a minute into the cotton fields, work previously done by platoons of weary workers. The solar collectors caught 40 percent of the available solar energy, exhibiting much better efficiency than earlier models in Pennsylvania had done.

Having devoted 7 years to experiments in solar-power engineering, spending a quarter of a million dollars—some of it his own money—in the process, Shuman now thought that solar technology had come into its own at last. He wrote in *Scientific American* in February 1914:

> *Sun power is now a fact and no longer in the "beautiful possibility" stage. . . . [It will have] a history something like aerial navigation. Up to twelve years ago it was a mere possibility and no practical man took it seriously. The Wrights made an "actual record" flight and thereafter developments were more rapid. We have made an "actual record" in sun power, and we also hope now for quick developments.*

Though the Maadi plant was small, its location (as Shuman and Ackermann had foreseen) guaranteed constant press coverage in Europe, especially England, where events in Egypt were followed avidly. The reaction in Europe and America to the news from Cairo about the solar power plant was positive. Former skeptics (including several at venerable American science journals) now praised Shuman's solar engine as "thoroughly practical in every way." Europe's colonial administrators heaped praise on Shuman's work, courting him in Paris, London, and Berlin on his visits there before the war. They spoke of the enormous economic benefit of using solar energy in underdeveloped Africa.

Frank Shuman and A. S. E. Ackermann thought that they were fulfilling the prediction of John Ericsson, made 38 years earlier, that

Egypt would become economically important because of the solar energy which fell on its land. Ericsson had asserted this in his *Contributions to the Centennial Exhibition* in 1876:

> *Due consideration cannot fail to convince us that the rapid exhaustion of the European coal fields will soon cause great changes with reference to international relations in favor of those countries which are in possession of continuous sun power. Upper Egypt, for instance, will, in the course of a few centuries, derive signal advantage and attain a high political position on account of her perpetual sunshine and the consequent command of unlimited motive force. The time will come when Europe must stop her mills for want of coal. Upper Egypt, then, with her never-ceasing sun power, will invite the European manufacturer to remove his machinery and erect his mills on the firm ground along the sides of the alluvial plain of the Nile, where an amount of motive power may be obtained many times greater that now employed by all the manufactories of Europe.*

A year after publication of Ackermann's *Scientific American* report, the Board of Regents of the Smithsonian Institution invited Ackermann to submit a monograph, *The Utilization of Solar Energy*, for the Smithsonian's annual report of 1915. Ackermann reviewed the whole history of solar power and commented extensively on the Shuman-Boys experiment in Egypt. This time, he focused more candidly on those aspects of the solar plant that required improvement:

> *Although the theoretical power value of the heat reaching the surface of the earth is no less than 5,000 horsepower per acre, it must not be thought that anything like this amount can be converted into mechanical power, any more than can all the heat of coal be converted into its theoretical equivalent of mechanical power. For example, the heat value of good coal is about 14,500 Btu per pound, equal to 12,760 horsepower hours per ton, but in fact the best result, even under test conditions, ever obtained from a ton of coal by means of a boiler and steam engine is only*

about 1,470 brake horsepower hours, or 11.5 per cent of the heat value, while in the case of a gas engine the corresponding figure is 25.5 per cent, and of a Diesel oil engine 31 per cent. The chief loss is in converting the steam into mechanical energy, and most of the loss is inevitable for thermodynamic reasons.

With this fact in mind, you will not be so surprised to learn that the best overall thermal efficiency obtained from the Shuman-Boys plant in Egypt was only 4.32 per cent, the chief reasons for this being so much less than 11.5 per cent [which is the best result ever obtained from a ton of coal] being that the steam pressure was so low, and that the best efficiency of the sun-heat absorber was only 40.1 per cent, compared with 75 per cent for the best coal-fired boiler.

But it has taken boilermakers many years to attain this efficiency, so that 40.1 per cent is not a bad result when the number of sun boilers that have been made is taken into account. Thermal efficiencies of engines are materially affected by the heat fall of the steam, just as the efficiencies of water turbines are affected by the height of the waterfall. The larger the fall in either case the better the efficiency.

Unfortunately, whatever its technical merits, the auspicious beginning for the Sun Power Company, Ltd., in Egypt did not lead to the growth Shuman so confidently expected. War was coming, and things were changing in Cairo. Frank Shuman's current and future prospects in Egypt were closely tied to those of his principal patron, Lord Kitchener, and Kitchener's earlier hopes for Egypt's economic development were no longer a top priority by 1914. His interest in constructive propositions like solar power had waned. With global conflict on the horizon, Kitchener had embarked on a dangerous plan to make radical political changes in Egypt.

The Maadi Plant Is Dismantled

The British consul's relations with Egypt's nominal ruler, Khedive Abbas Hilmi II, never good, were deteriorating. Egypt's king was a self-indulgent playboy, but he was no fool. He was careful to cultivate an outward popularity with the Egyptian person-on-the-street, serving as a rallying point for Egyptian nationalists. Though most of the latter despised him and would admit privately that if Egypt ever became independent, the monarchy would be the second institution to be thrown out (right after the British) everyone realized that having even a nominal soverign was a potent symbol of independence.

Kitchener knew the private Abbas Hilmi, whom he loathed. From that moment two decades earlier when he had humiliated him on the battlefield, Kitchener could not bring himself to take the Egyptian ruler seriously. He was just "that wicked little khedive"—a clown and an embarrassment. Kitchener complained to Lady Salisbury, his favorite correspondent in England, that he often felt, in having to deal with the khedive at all, that he was playing a part in a comic opera rather than conducting "serious government." By 1913, Kitchener had determined that serious government meant a government free of Abbas Hilmi, and he began to conspire with his superiors in London to rid Egypt of him permanently.

Before he left Cairo in 1914 for his summer vacation in England (he was informed in early June that the king was going to make him a gift of an earldom in recognition of his services to the Empire), Kitchener set into motion a plan whereby Great Britain would formally annex Egypt. The timing for this plan could not have been worse. Kitchener reached Dover on June 23, 1914, 5 days before the assassination of Archduke Francis Ferdinand, the heir to the throne of Austria-Hungary, providing the spark that ignited the Great War. "One month later," writes Philip Magnus in *Kitchener, Portrait of an Imperialist*, "the throne of human reason itself was temporarily overwhelmed by the mightiest tornado which had ever risen to the surface of men's minds from some remote atavistic depths to sear and blast humanity." It was exactly 2 months after the completion of the last trials of the Maadi solar plant, putting the solar machinery into regular and permanent operation.

Abbas Hilmi, himself on his annual summer tour in Europe, got wind of the plot Kitchener had launched against him and announced his intention to go straight to London to appeal, king to king, to

George V. Kitchener acted immediately to block the visit. He met with his sovereign at the end of June and the next day the khedive was informed by the British ambassador in Paris that if he came to England, the king would decline to see him. In the event, neither the khedive nor Kitchener would ever set foot in Egypt again. Neither, for that matter, would Frank Shuman.

On November 5, 1914, war broke out between Great Britain and Turkey and a British protectorate was declared over Egypt. Abbas Hilmi, by then in Constantinople to visit the Sultan of Turkey, retorted by condemning the British action. That was the last straw for politicians in London, who immediately clamored for his removal. The khedive was soon formally deposed. He sat out the war in Constantinople.

Kitchener remained in England, where, as Britain's most prominent military man at a time when military men were seen as urgently needed to save the nation, he was the toast of the glittering London social season. During the weekends, he usually went to Broome Park near Canterbury, the magnificent 500-acre, seventeenth-century estate he had purchased in 1910 (for a mere 14,000 pounds sterling) and which he was busy restoring for his eventual retirement. He told King George in July that he would like to go to India as Viceroy. The king seemed disposed to give him the job, a plum Kitchener had long coveted.

It was not to be. England's most famous soldier, now 64, was needed to give the war effort a high profile. In August, Prime Minister Asquith appointed him Secretary of State for War, the first serving soldier to sit in a British cabinet since 1660. Asquith had reluctantly offered him the job only because of Asquith's hope that Kitchener's enormous popularity would bolster his own, which was fading.

Though many British politicians expected a quick victory, Kitchener correctly predicted how bloody and protracted the Great War would really be. He took on his new duties with reluctance—the prospect of many happy years at Broome had captured his imagination and he longed to spend his time as a country squire. Kitchener would not survive the war. On June 5, 1916, he drowned in the cold waters of the North Sea, off the Orkney Islands. The Royal Navy cruiser HMS *Hampshire*, on which he was traveling to Russia on a diplomatic mission, was cut in half by a German mine (one laid by a German submarine a few days before). The *Hampshire* sank within 10 minutes, with most of her officers and crew. Kitchener's body was never found.

Kitchener's permanent absence from Egypt was only the first of many disasters that were to befall the Sun Power Company's installation at Maadi as a result of World War I. The fighting in Europe that followed the assassination of Archduke Francis Ferdinand spread to Europe's colonial territories, and the northern regions of Africa were soon engulfed by war. Shuman's solar irrigation plant was dismantled for parts and much-needed scrap metal. The engineers associated with the project returned to Europe and America to perform war-related tasks.

Whatever the strengths and versatility of solar energy, even its most vocal advocates conceded that solar-powered machines were not suited to propelling ships or land vehicles—and irrigation projects, in Egypt or elsewhere, suddenly assumed a low priority in a world girding for battle. At a time when Europe itself was threatened, the driving force behind solar-powered engineering had disappeared. By late 1914 and early 1915, most politicians and many scientists in Britain would have deemed solar-power experiments to be a frivolous, even treasonous distraction. Under these conditions, Shuman and his engineers could not continue their work in Egypt.

Had war needs alone stalled Frank Shuman's grand solar experiment in Egypt, solar research might well have recovered after the armistice. Indeed, while taking what he thought would be a temporary sabbatical from Egypt, Shuman worked on an insanely ambitious proposal to build more than 20,000 square miles of reflectors in the Sahara designed to give the world "270 million horsepower per year in perpetuity," which was, he had calculated, the equivalent of the energy content of all the fuel mined worldwide in 1909. Shuman knew that the war would eventually burn itself out, and he hoped to pick up where he had left off with still greater projects. But by the middle of the second decade of the twentieth century, there were far more ominous threats taking form that were only indirectly related to the Great War. The most important had quietly been on the scene for nearly half a century. It was called petroleum.

The Age of Oil

By 1915, the whole of the American and European economies were dependent on coal: in the steel industry, for shipping, electricity generation, and for a host of complex chemical processes that provided raw materials to industries as diverse as pharmaceuticals and munitions. Coal was also the basic heating and cooking fuel in most homes and factories. Threats to coal production were seen by the public and governments alike as tantamount to acts of war (in much the same way as threats to oil production would be perceived 90 years later).

Britain's Great Coal Strike, which began in January 1912 in the coal fields of Wales, had virtually crippled the British economy. When more than a million miners from Glasgow to Newcastle put down their picks for higher wages, almost everyone in the nation turned against them, even labor unionists who might have been expected to support them. Far from generating sympathy among the working class (as the earlier transportation strike had done in the autumn of 1911), the coal strike created bitter resentments on all sides of the dispute.

Because the commitment to coal was so entrenched, far from mobilizing Europeans to explore new sources of energy, labor disputes and war had the opposite effect: Europe's rulers hunkered down and sought to control all available production of the only energy source that was absolutely tried and true. And yet, while the Great War strengthened world dependence on coal in the short term, it also hastened a longer-term trend that many had seen starting decades earlier—the ascendancy of petroleum as the world's primary fuel source for transportation.

From the evidence that Frank Shuman and others were accumulating in the first decade or two of the twentieth century, it certainly seemed that solar-powered steam generation might have given coal a run for its money after the conclusion of the war, at least in certain parts of the world and for certain applications, like irrigation. But solar power could not compete with the potent new fuel that was coming to the fore as a result of the global conflict. One of the earliest lessons of the new century was the importance of oil as an energy source, especially in war. It conferred many advantages over coal in mobilizing armies and navies.

Petroleum was no longer the novel substance used for medical nostrums that it had been in the days of Napoleon III's first Paris exposition, when it was displayed alongside Mouchot's solar steam engine. Petroleum was now recognized as an important energy source, though rarely in its early years with the conversion of heat into mechanical energy. Oil's original strength was seen in its use providing artificial lighting.

In the United States, petroleum's rise had begun in Frank Shuman's native Pennsylvania. There, "Colonel" Edwin Drake drilled some of the first oil wells in a systematic way, starting in 1859. Eleven years later, in 1870, John D. Rockefeller formed the Standard Oil Company mainly from Pennsylvania-based production. Shell Oil was set up as a Far East trading company in 1878, combining with Royal Dutch Petroleum in 1907. What became Texaco was created in 1902, and BP/Amoco's original predecessor was formed in 1909 by the legendary William Knox D'Arcy. By 1900, long before the advent of serious motorized transport based on petroleum derivatives, world oil production already averaged 400,000 barrels a day. American oil fields dominated this output with 174,000 barrels a day, but other major producers included Russia, Canada, Mexico, and Argentina.

By 1900, Rudolf Diesel's engine was already a mature invention with 2 decades of improvements behind it, though coal and steam were still dominant in rail and sea transportation as well as in industry. Petroleum refineries often dumped volatile elements like gasoline into rivers as a waste product, with potentially deadly effect, because there was no use for these explosive "cuts" of the refining process. Kerosene, the most prized petroleum product, dominated the early years of refining because it could turn night into day for millions by providing cheap illumination. But kerosene was also volatile and potentially dangerous to use. As a result of Thomas Edison's work with filaments and dynamos, kerosene was gradually being supplanted as a source of illumination by electricity, which produced better-quality lighting without smoke or risk of flame.

By 1910, fuel oil was beginning to make some inroads in replacing coal in factories and in powering locomotives and ships, but it was the First World War that would raise oil to be the principal source of energy for transport, largely as the result of the political force of will of one man. Winston Churchill, Great Britain's 39-year-old First Lord of the Admiralty, was among the first to recognize petroleum's military significance as a flexible and powerful alternative to coal, and he was determined to make the British navy an oil-fired fleet.

The British Navy Turns to Oil

In Daniel Yergin's prologue to *The Prize,* his riveting account of the rise and expansion of the oil industry, he argues that it was Churchill, more than any other single individual, who inaugurated the "age of petroleum." When he was put in charge of England's navy, Churchill recognized at once, as few others in the Admiralty did, that the age of coal, at least as a source of propulsion for warships, was over. That was the main conclusion of an analysis he had made of the German fleet, England's likely adversary in the coming war. Ships fueled by petroleum were faster and more maneuverable than ships with coal-fired boilers and clumsy steam engines or turbines. To face the German challenge on the high seas, Churchill was determined to have an oil-powered fleet at England's disposal, even at the cost of gutting the steam-powered engines in existing vessels and replacing them, a colossal expense even for a wealthy nation.

Churchill's plan created an outcry within the British government, for it seemed on its face a very risky strategy: Britain had a modest indigenous coal supply safely within her own borders, mainly in Wales, but (in the years before North Sea discoveries) no oil reserves whatsoever. To make the Royal Navy, the backbone of England's military, completely dependent on foreign oil seemed to most military analysts to be a dangerous folly. Even Churchill recognized, that "to commit the Navy irrevocably to oil was indeed 'to take arms against a sea of troubles.'"

"But the strategic benefits," writes Yergin, "greater speed and more efficient use of manpower, were so obvious to him that he did not dally. He decided that Britain would have to base its 'naval supremacy upon oil' and, thereupon, committed himself, with all his driving energy and enthusiasm, to achieving that objective."

Earlier efforts to convince British officials to use fuel oil in warships had not been successful. A fuel oil test on HMS *Hannibal* was described by the *London Illustrated News* as a successful experiment "in terms of locomotion, but has the drawback that ships using oil emit enormous volumes of dense black smoke—very picturesque but by no means convenient or pleasant." As early as 1899, Marcus Samuel, the founder of Shell Oil, had begun lobbying the British government to convert the Royal Navy from a coal-burning to an oil-burning fleet. Samuel convinced the First Sea Lord, Admiral John Arbuthnot Fisher, that oil would give the British Navy an advantage over the coal-burning German fleet. Fisher, in turn, may have helped to sell Churchill, who by then headed the Admiralty, on the idea. While Frank Shuman and A. S. E. Ackermann had been plotting their Egyptian campaign in 1912, Churchill was already issuing orders that the British fleet be converted.

Beyond bringing the Royal Navy's vessels into dry dock to be outfitted with British versions of the Diesel engine, Churchill's more daunting task was to find a secure supply of petroleum for his nation that would last for decades. Churchill persuaded the government to take a 51 percent stake in Anglo-Persian Oil Company (today, many mergers and consolidations later, called BP/Amoco) in an effort to bring Mideast oil under permanent British control. Anglo-Persian's concessions in Iran, though outside the British Empire, were sufficiently close to entrenched British holdings in Burma and India so that British troops could speedily be dispatched there in the event of a crisis or threat. By having the government take a big stake in the firm, as it had in the Suez Canal, Churchill also guaranteed, if tacitly, that Anglo-Persian would always be financially solvent and remain firmly under British control. Though outwardly a private firm, it was in fact the first state oil company, one created for political and strategic rather than economic reasons.

The availability of oil for the allies in World War I was likely one of the most important factors in their victory. When submarines were endangering Western Europe, Marshall Foch observed that "we must have oil or we shall lose the war." He had demonstrated the value of oil when, challenged to call up more French troops for the Battle of the Marne, he commandeered civilian taxis to take thousands of soldiers to the front by "cab."

In geopolitical terms, oil discoveries made the United States the most powerful nation in the world, though the country did not yet realize it. Everywhere, the balance of power changed as a result of the first oil war. The Allies, Lord Curzon said later, "floated to victory on a wave of oil." The chief of the German general staff, Erich Ludendorff, later blamed Germany's defeat on its lack of oil. The French foreign minister, Aristide Briand, noted, "In our day, petroleum makes foreign policy." Georges Clemenceau said that oil was as necessary as blood.

Oil is a much more flexible fuel than coal and holds within it a denser concentration of energy. Because it can flow, it is more easily transported. It is cleaner when burned. Much of the time, it was cheaper than coal. Three barrels of oil have the heating capacity of 1 ton of coal and at the prevailing prices early in the century, the oil usually cost only half as much as coal. Because of oil, the number of men tending the furnaces on a steamship could be reduced from 100 to four. Loading a ship with coal had taken 100 men toiling for a week; now one man in one day could load a ship with enough bunker fuel to cross an ocean.

The industrial as well as the consumer possibilities oil held out appeared to be endless. Oil meant that the machine age could move beyond the hulking steam engines that dominated the nineteenth century. Smaller machines and consumer machines, such as those being developed by Henry Ford, could have practical and inexpensive daily application. Small, gasoline-powered motors supplanted the compact heat engines developed by Ericsson to power fans and other small, portable appliances.

Within a year of the huge Spindletop discovery in Texas, oil was being put to all kinds of new industrial and consumer uses, and the

U.S. economy, too, began visibly to switch from coal and to oil. American and European society was in the process of being modernized, industrialized, mechanized, and electrified.

A New Mistrust of Technology

Though wars often spur certain areas of research, the years immediately after Francis Ferdinand's assassination were not conducive for benign research of any kind, nor for the proliferation of technologies, like Shuman's, that brought no obvious advantage to the prosecution of the war. In fact, those energies that had powered the West's industrialization were now brought to bear in the most destructive war the world had ever seen. The Great War shattered the public's belief in the benign power of technology, for it was technology, like a flywheel spinning too fast, that now threatened to tear Europe apart.

It may not be possible for people living on the cusp of the twenty-first century to understand the almost mystical reverence people living in the divide between the nineteenth and the twentieth had felt for technological progress before the First World War. The Shumans and Ericssons of the Victorian and Edwardian eras were viewed as benefactors and their creations as benevolent gifts to humanity. The development of the steam engine represented the first sure steps in humankind's eventual, inevitable, triumph over nature, they thought. Before the development of quantum theory, most well-read people on both sides of the Atlantic thought that almost everything that could be understood in science had been revealed and that little remained but to put all this newfound knowledge to practical use, bettering lives. There had always been dissenters to this view—Mary Shelley, the poet Carlyle, and the Luddites come immediately to mind—but by and large scientists and engineers were viewed by most people as heroes before World War I.

With the new horrors of industrialized warfare, engineers could no longer be hailed as modern-day saviors, their works greeted as panaceas, and their efforts charged with humanity's confidence. Belief in technology was a hallmark of all the nineteenth-century solar pioneers—a faith that, no matter how intractable the problem, there were answers to society's ills that science and technology would reveal. Now, mustard gas and mechanized warfare obliterated that innocence. By the end of the war, Bertrand Russell was able to write that "whether, in the end, science will prove to have been a blessing or a curse to mankind is to my mind still a doubtful question."

A new attitude toward technology more reflective of Russell's view than of Shuman's became entrenched from 1915 on. Moreover, American and European commitment to a petroleum economy was irreversible, consigning solar power to a backwater in which it would languish until mid-century, when interest in the photoelectric effect supplanted earlier interest in mechanical devices powered by solar-derived steam or hot water.

In the meantime, oil went on to dominate twentieth-century politics and business in a way that overshadowed even coal's domination of the nineteenth century. By the end of the First World War, Britain's control of the Anglo-Persian Oil Company was ironclad. Within the next few years, hundreds of new oil and gas fields were discovered around the world—many in arid, sunny areas where coal was difficult to obtain, often precisely the locations earmarked by Frank Shuman as most suitable for solar-power plants. With automobile sales booming and oil and gas in such surplus that it was available practically at cost, concerns about energy shortages evaporated, and the prospects for solar power were eclipsed. Who needed solar power when oil power was practically free?

Shuman's giant solar machine on the banks of the Nile, paradoxically, was the *culmination,* not the beginning, of a technological trend—a diverting trip down an alley that turned out to be a blind, at least until the final decades of the twentieth century, when there was a major revival of interest in solar-generated steam—this time to run electric dynamos.

That is not to deny the credit that Shuman is due: Shuman's invention worked, and it worked on the cheap. By measures technical and economic, it was a success. There were kinks to be worked out, and later models, had there been any, would surely have been cheaper and more efficient than the device at Maadi. In the meantime, the patient died, as the old saw goes. Maadi was the end of the road.

Though it became clear in the first two decades of the twentieth century that gasoline- or diesel-based engines were incontestably the machines of choice for powering ships, airplanes, and automobiles, there were many large applications where they were not economic—in manufacturing electricity, for example. At the turn of the twenty-first century, the world still produced much of its electricity using steam turbo-generators that differed little from those in use in Shuman's lifetime. And yet, the generation of mechanical energy through solar power was too young a branch of engineering to follow a similar track. It was unable to establish a foothold, under these conditions, that would lead to a long-term lease on the public imagination in the new age of petroleum. And so Shuman and his predecessors, and all their inventions, slipped into the shadows of a profound and long-lasting obscurity.

Converting Solar Energy to Electricity

Whether or not some new set of circumstances after the Great War along with Shuman's device could have sparked the commercial success that solar power needed, we will never know. It does not seem likely. But while the mechanical dimension of solar-power technology got sidetracked, another path to capturing the sun's power—direct conversion into electricity—was making headway. This technology proved so powerful that by the 1950s photovoltaic (PV) cells developed at Bell Telephone Labs in Murray Hill, New Jersey, had eclipsed efforts to harness the sun's energy by mechanical means, efforts not revived until the 1980s. These "solar cells" (as they were dubbed by *Life* magazine) could transform solar energy directly into electricity, with no moving parts.

Toward the end of his career, Augustin Mouchot had experimented with solar generation of electricity, but he was using the action of solar heat on dissimilar metals to make an electric current. In photovoltaic cells, it is light itself, rather than heat, that generates electricity. The ability of light to produce this effect had been known since 1839, when Edmund Becquerel, a French experimental physicist from a family that produced four generations of scientists, discovered that sunlight could generate a flow of electrons in certain materials. Becquerel's experiments were conducted in liquids. It took 50 years before Charles Fritts, an American inventor, made the first primitive solar cells out of solid material. These thin wafers, each about the size of a quarter, were made from selenium covered with a transparent gold film. When sunlight struck the coated wafers, a current was generated that was "continuous, constant, and of considerable electromotive force," according to Fritts's laboratory notes.

Though Fritts believed that at least half the light hitting the surface of these cells could be converted into electrical energy, his first trials converted something less than 1 percent of the light energy reaching the selenium into electricity, a level of performance that would not be much improved upon until the Bell Lab experiments in the 1950s. The story of photovoltaics has been one of consistent and steady progress since then.

The End of an Era

Frank Shuman was less obviously a casualty of the Great War than Lord Kitchener was. Apostles rarely retire, and Shuman never did. He remained in the United States after his final voyage on the SS *Mauretania*. He sat out the hostilities at his roomy compound in Tacony, working on inventions to improve submarines and a revolutionary proposal to build a huge land battleship 200 feet high. He became active in the Franklin Institute and the Engineers' Club of Philadelphia. He spent long afternoons at the Torresdale Golf Club. In 1917, he perfected the submarine he had earlier taken (in concept) to the U.S. Navy, a revolutionary vessel that was to be propelled by liquid oxygen, which would also provide breathable air for the crew. He again offered the device free of charge to the government as his contribution to the war effort. It was again rejected, but after the appearance of a similar invention in Germany, a patent was promptly granted. Though they were not buying his inventions, Navy officials were glad to see Shuman comfortably retired and not making trips to Germany with such potentially dangerous ideas in his head.

In 1939, long after Shuman's death, family members recounted for a Philadelphia newspaper that the U.S. Navy's rebuff of his submarine work had been more devastating to Shuman than the failure of the solar project in Egypt—the latter disaster being comprehensible in the context of the exigencies of the war in a way the former was not. To the end, in spite of evidence to the contrary that even he must have attached some credence to, Shuman appears to have been certain that solar steam engines would experience a renaissance after the end of the war. Shuman's successes in life, as well as his fortune, had been based on his trust in his own inventive instincts, and he trusted them now to the end.

When the U.S. submarine *Squalus* sank in June 1939, taking 26 lives to the bottom of the ocean, the *Philadelphia Evening Bulletin* resurrected the story of Shuman's patent to provide liquid oxygen tanks in submarines as a source of breathable air. By then, the huge family compound on Disston Street had been dismantled and sold. Shuman's widow was living in a comfortable apartment in central Philadelphia. She still had copies of the now-yellowing patent application and brought them out for an inquisitive reporter.

"They haunt me, those sailors who gasped and died for lack of oxygen on the ocean bottom, because right here in this drawer are my late husband's plans for an invention which would have saved every one of them," Shuman's widow said. The Navy disagreed. Submarine experts at the Philadelphia Navy Yard told the paper that they doubted that the Shuman invention would have been the salvation of the men who died in the *Squalus*, or the 99 who died in the British submarine *Thetis*, or 71 who died in the French submarine *Phenix*—all recent submarine disasters. Shuman's own appeals to the government to adopt his invention had grown more strident when the U.S. submarine *F-1* was rammed and sunk by the *F-3* with a loss of 19 lives during maneuvers off San Diego in 1917. He insisted to his death that a kind of institutional conservatism in the U.S. Navy, the same conservatism that had caused admirals to cling to sails for a quarter-century after the Navy's own engineers had developed faster steam warships (the same conservatism, in fact, that had enraged and frustrated John Ericsson), was all that had blocked recognition of his invention.

"Some day they'll have to adopt something like it if they continue to use submarines," he said. And in one of his last published articles he declared that "the truth is that, for all its deadliness, the submarine is a very crude piece of machinery."

Frank Shuman died of a heart attack on April 28, 1918, at his home on the grounds of his inventor's compound in Tacony. He was 56 years old, a relatively young man even in that era of shorter life expectancies. Seven months later, on November 11, 1918, the German emperor surrendered, bringing an end to the Great War.

A. S. E. Ackermann survived not only the First but also the Second World War. He was called back to England from Cairo at the outbreak of hostilities in 1914 to serve in the Ministry of Munitions. Later, he was seconded as a consultant to the National Aircraft Factories. He resumed his Westminster consulting practice in 1919, a year after Shuman's death, but without the spectacular success he had enjoyed before the war. He wrote a widely read and often reprinted book, *Popular Fallacies*, and became an early broadcaster on technical and engineering matters for the British Broadcasting Corporation. In an odd departure from his previous interests, he became the world's leading expert on the properties of clay, authoring a number of monographs on the subject. He died at the age of 83 in April 1951, having never taken up again any of his solar studies since his departure from Egypt 35 years earlier.

New Developments

Granting that the First World War changed world energy dynamics with a decisive shift to petroleum in the first decades of the twentieth century, how is it that solar-power technologies did not take firmer root in France, Great Britain, and the United States in the last half of the nineteenth century, when it was clear that (at least under certain conditions) solar power could compete economically with coal? What-if questions rarely provide answers, though they often give rise to interesting speculation.

In the continuing revolution in sources of power for machines from 1780 on, what is more remarkable than solar's failure is the sheer number of energy-related innovations in the last few decades before 1900. Technological breakthroughs made engines more efficient—the improvements by Parsons and de Laval in steam turbines in the 1880s; by Daimler and Benz and Diesel in the 1890s in the development of the internal-combustion engine; and by Edison and European engineers such as Siemens and Ferranti in the generation and application of electric power in the 1880s—all these were revolutionary in the sense that they sowed the seeds of technological progress for decades into the future.

The fact that, early in the twenty-first century, we are still producing much of the world's electricity using coal-fired steam boilers and that our vehicles and ships are still powered by internal-combustion engines, demonstrates the extraordinary staying power of these innovations. The most revolutionary of the pre-1900 developments was surely the wide-scale use of electricity. Electricity provided industrial mobility. Before electricity, heavy industry depended completely on coal mines, and all big plants had to be built near them. Electricity could be generated from a variety of diverse sources, not just coal, and the electrical energy itself could be moved great distances in copper wires with little energy loss. The fragile new solar technologies had to compete with this onslaught.

It has often been said that technology tends to create its own environment and set of conditions. As the technology increases in power and scale, the tendency is to fit new situations into the existing technology rather than to find new technologies that better suit the new situations. Winston Churchill's achievement in reengineering the Royal Navy from coal to oil is all the more remarkable in light of that fact. The odds against his success, in an era of entrenched coal technology, were as great as those against the success of electrical cars today, in an era of entrenched petroleum technology. Coal, oil, kerosene, and other combustible fuels in 1914, now abetted by electrical power, were becoming too entrenched in Europe to make room for the wayward and unproven solar interloper.

What if Shuman had built the Maadi plant in 1893 instead of 1913? Would a 20-year head start have given solar steam technology the opportunity to find a small, permanent niche in tropical regions where coal was prohibitively expensive? Quite apart from Shuman's disastrous timing (his was not the first scientific or technological project, nor the last, to be crushed by a war), other factors were at work in bringing further development of solar-power generation using steam to a grinding halt in 1914. It would not recover until the early 1980s.

Shuman and Ackermann worked at a unique time, in the transition period between the nineteenth and the twentieth centuries. The emerging solar technology of the era was inextricably linked to the economic, political, and cultural forces of the prewar years. It is difficult to disentangle science and technological progress from the social and economic factors with which it is entwined; indeed, in some sense, they are the same thing. It is often not clear whether economics drives technology or the other way around, though it is almost always true that brilliant technologies that are not economic rarely succeed. It has been wisely said, too, that while a brilliant scientist must have new ideas, brilliant engineers are those whose conceptions rely on as few new ideas as possible. As Shuman and Ackermann took such pains to demonstrate, at least at the microeconomic level, their solar-powered engine did succeed, at least in certain locales, in generating cost savings over any reasonable period of time. The front-loaded investment diluted that fact, especially in an era (a war) when planning horizons were curtailed. But in the end, theirs was a strange contraption perhaps too full of new ideas.

Those areas where Shuman's solar machines would have been most effective were precisely the places where the wealthy and influential did *not* live. A technology that provides for economic progress in Egyptian or American deserts may be of little interest to people in London, Paris, New York, or Chicago. Shuman's niche was too narrow, the impact of his inventions too parochial, for them to have had the global impact of Edison's lightbulb or Marconi's radio. Under different circumstances, Shuman himself might have foreseen this—after all, one of his early financial successes, Saftee-Glass, was founded on the almost universal application of his invention. Large inventions require large markets, and solar's market in 1910 was simply too small.

Shuman can be forgiven his optimism. The economic conditions Frank Shuman was familiar with were above all expansionist—

expansion of population, of manufacturing, of trade, and of scientific knowledge. In his time, these increases often seemed to him to have no foreseeable limit, and they were taken to herald the achievement of a kind of universal progress. Yet, far from producing peaceful progress, the end of the nineteenth century ushered in a transitional period of upheaval and violence, resulting in a distrust of technology.

The pace of application of scientific discovery was speeding up throughout Shuman's lifetime. For the first time, people saw discoveries in pure science made in one year, with no obvious practical effect, producing commercially useful devices in the marketplace the next year. It was hardly a time that favored wallflowers, and so Shuman charged ahead.

Frank Shuman was remarkably prescient in picking the field to which he would devote his greatest personal efforts, after making his fortune mainly in patenting safety devices. He correctly understood that the greatest achievements of the physical sciences of the late nineteenth century—the doctrine of the conservation of energy and the interchangeability of the various forms of energy, the sciences of thermodynamics and electrodynamics—drew their inspiration from the study of practical sources of power and arose from the new need to transport people and things, and to communicate over long distances quickly. He foresaw the limitations of resources like coal, which are nonrenewable, though his timing may have been off by a couple of centuries. He was concerned about environmental effects in an era when that perspective was almost unknown. If he did not foresee the rise of the internal-combustion engine, he can be forgiven that, for so much of the industry of the nineteenth century depended on the development of, and the continued improvement of, the steam engine. Exogenous developments are rarely foreseen even by brilliant men.

Science in the nineteenth century was largely the product of individual efforts of men of genius. Science in the twentieth century, beginning with the convergence of technology and war that characterized the Great War, quickly evolved into a highly organized new profession closely linked with industry and government. After World War I, governments and large corporations set up special institutes or study groups to investigate on a mass level various problems—the return on investment in research, the reasons for lags in the application of scientific discoveries, optimum investment in basic and applied research, and other issues—that had been dealt with previously by individuals or small groups of inventors. After World War II, it was the rare global company that did not have a research-and-development division, usually many times larger than Edison's inventor's village, whose mandate was to delve into theoretical science and to find there practical applications for profit.

It is hard today to grasp the uncoordinated and amateur character of nineteenth-century science, with little formal teaching and without research laboratories or research funds and with so much left simply to chance. It is almost as difficult in an age of vast engineering and chemical factories, each with its own research and development department, to recall the intimate traditional and practical bent of the old workshops and forges from which so many inventions we still use are descended. Shuman was a tinkerer and lone inventor, a gifted eccentric, in the finest of that tradition.

New Possibilities

Why hasn't solar-power technology taken off in the century since Shuman built his great machine outside Cairo? Traditionally, concerns about cost and reliability are cited, though politics and tax policy have played their part. Electricity shortages in 2001 and skyrocketing electricity rates in California, concerns about rising gas prices, and scarcity of heating oil in New England remind us that no story about energy, even solar energy, is ever merely of historical interest. Steady decreases in price and more sophisticated storage mechanisms are making solar power more competitive, especially in the United States. (That nation alone among the industrialized nations of the world, let us not forget, is blessed with copious sunlight; one can forgive Germans and Scandinavians for not thinking their economic futures will be hitched to solar power!)

Since 1980, solar power has grown 15 to 20 percent a year in the United States (albeit from a tiny base), a phenomenal rate of growth in spite of availability of the cheapest grid electricity in the world, with prices from 6 cents to 15 cents a kilowatt-hour (compared to 20 to 50 cents a kilowatt-hour elsewhere in the world). And yet, by and large, solar energy continues to elude efforts to tame it, to popularize it, and to help it become decisively competitive with its alternatives. The lament of John Ericsson is as timely in its general import in the twenty-first century as it was in the nineteenth: "Although the heat is obtained for nothing, so expensive and complex is the concentration apparatus that solar steam is many times more costly than steam produced by burning coal." As Shuman himself so wisely observed, "This is the rock upon which, thus far, all sun-power propositions have been wrecked."

Sunlight is free and strikes the earth in spectacular multiples of any conceivable total world demand for energy, but harnessing the sun to do useful work involves substantial investment. That hardly deterred Ericsson, who had the stiff resolve of a Swede coupled with the ingenuity of a sharp-elbowed immigrant New Yorker. Nor did it deter Shuman, who never underestimated his ability to shift the tide of economics with new inventions. It is entirely possible that both men were right and that their only error was one of timing. With time, the investment required to convert solar heat into useful energy is decreasing. Sober and prudent men reasonably expect that solar technologies, including solar steam technology, will become even more cost-effective in the future.

One need not ask why men like Frank Shuman, John Ericsson, John Adams, and Augustin Mouchot were willing to take enormous personal risks to harness the sun's energy. No matter how one chooses to define it, the prize that eluded them is staggering. The work that can be derived from the sun's radiation is almost beyond calculation. When measured in units of horsepower or megawatts, the numbers are followed by so many zeros the mind scarcely can grasp them. Each day our planet receives as sunlight about 200,000 times the total world electrical-generating capacity. In 45 minutes, the United States receives more energy in the form of solar rays than it uses in a year from all forms of fuel. If the solar energy that reaches just 1 percent of the Sahara desert could be captured and converted into electricity (as Shuman once calculated), it would meet all the electrical energy needs of all the nations of the world. Alas, this bounty of solar energy is in the form of light and heat, and heat alone is of little use. Converting heat to useful energy, and doing so cheaply, is where economic genius has been required.

But progress is being made. On a more down-home scale, the amount of solar energy falling on 1 square mile of Midwestern farmland at noon on a cloudless day is just over 1750 megawatts. If only 20 percent of that could be converted into electricity (easily attainable with today's solar cell technology), one would have enough power to serve a community of 35,000 typical U.S. residential customers—even assuming that clouds and bad weather reduced collection time to an average of only 4 hours daily. The solar energy falling on just 28 square feet, if it were constant and could be converted at 100 percent, would supply all the energy needs of an average household; factoring in 20 percent conversion and only 4 hours of sunlight per day, an area the size of a tennis court could easily supply all the energy needs of two or three American households on a continuous basis.

The other great lesson that Shuman drew from his own and his predecessors' experiments was that solar energy is indigenous energy, as local as his own small neighborhood of Tacony, as neighborly as a town meeting in New England, as independent as pioneers on a wagon train crossing the sunlit West. It keeps reappearing on the American agenda. When heating oil runs short or electricity browns out, citizens write their legislators and newspaper editors about power from the sun. No one wants a coal-burning power plant in his or her backyard, or the cooling dome of a nuclear reactor visible on the horizon. Politics will certainly play as important a role in solar's future as economics.

The sun, meanwhile, rises benignly every day—and shines. It will likely continue to do so for tens of millions of years. Solar energy endures. It will surely seize the imagination of some future engineer as firmly as it gripped the mind of that hardy American inventor on the banks of the Nile nearly a century ago.

Background Information

[Books referenced only by author's name refer to volumes in the bibliography, where full citations will be found.]

Chapter 1 Philadelphia's Solar Wizard

- For details of Cunard's SS *Mauretania*—her specifications, facts relating to the provisioning of the great ocean liner at the beginning of the twentieth century—as well as colorful descriptions of New York in that era, I am indebted to Brinnin's encyclopedic and beautifully illustrated *Beau Voyage.*
- A useful (though rather stark) compilation of biographical facts and dates relating to Shuman's life is found in his longish entry, "Frank Shuman," in *The National Cyclopaedia of American Biography,* Vol. XIX (New York: James T. White & Company, 1926), a copy of which was kindly provided me by the Free Library of Philadelphia.
- I have relied heavily on Silcox's two published works on the history of Tacony and the Disston industrial empire (see bibliography) in painting my picture of Shuman and his Tacony inventor's compound at the turn of the century. Details of the Tacony compound, Frank Shuman's life and work before Egypt, and especially his early inventions, are also derived from interviews with Professor Silcox, who gave me a wonderful sense of the vibrant life of the Philadelphia suburb at that time. Tape recordings Silcox made in the 1980s with Shuman's aging children, which he shared with me, were also invaluable, as well as his own recollections, shared by telephone and email, of those interviews.
- Mrs. Sue Shuman Widing, the granddaughter of Frank Shuman's brother, Constantine, was generous in sharing family stories and in

helping me understand how the large Shuman family was supportive of Frank Shuman's efforts as an inventor.

- Ms. Siobhan Gephart, executive director of the Historical Society of Tacony, provided background information about Shuman and Tacony from the archives of the society.
- Frank Shuman described his Tacony experiments in "Power from Sunshine, A Pioneer Solar Power Plant" in a long, bylined article illustrated with photographs in *Scientific American*, September 30, 1911.
- The discussion of converting the Sahara into an inland ocean is in "A Plan for Converting the Sahara Desert into a Sea: What Would Happen If the French Flooded the Great Desert," by G. A. Thompson in *Scientific American*, August 10, 1911.
- Robert H. Thurston's pessimistic outlook for coal production, as well as his thorough historical review of solar-power engineering up to the turn of the century, is in "Utilizing the Sun's Energy," in the *Annual Report of the Board of Regents of the Smithsonian Institution* for the Year Ending June 30, 1901, pp. 263–270.
- The original letter from Shuman to Judge Thomas W. South, dated August 13, 1907, on Simplex Concrete Piling Co. stationery, is in the collection of the Historical Society of Tacony.

Other articles by or about Shuman that were helpful in researching this chapter include the following:

- "Feasibility of Utilizing Power from the Sun," by Frank Shuman in *Scientific American*, Vol. 110 (February 28, 1914), p. 179.
- "The Generation of Mechanical Power by the Absorption of the Sun's Rays," by Frank Shuman in *Mechanical Engineering*, Vol. 33 (December 19, 1911), p. 1.
- "Sun Power Plants Not Visionary," by Frank Shuman in *Scientific American*, Vol. 110 (June 27, 1914), p. 60.
- "Solar Power," by Frank Shuman in *Scientific American*, Vol. 71 (February 4, 1911), p. 78.
- "The Most Rational Source of Power: Tapping the Sun's Radiant Energy Directly," by Frank Shuman in *Scientific American*, Vol. 109 (November 1, 1913), p. 350.
- "Power from Sunshine, A Pioneer Solar Power Plant," by Frank Shuman in *Scientific American*, Vol. 105 (September 30, 1911), pp. 291–292.

- The discussion of total horsepower available in Europe for commercial applications at various times before, during, and at the end of the industrial revolution is adapted from Hills's erudite *Power from Steam*.

- In this, as in so many of the chapters that follow, I am indebted to Ken Butti and John Perlin for facts and material on Shuman, Ericsson, Mouchot, Tellier, Eneas, Willsie, and Boyle, as published in their encyclopedic *A Golden Thread,* perhaps the most readable book on the history of solar power.

Chapter 2 Shuman's Inspirations: Solar Power in Ancient Greece and in Europe in the Middle Ages

- The 1992 British investigation of the physics of Archimedes and the burning mirrors of Syracuse is documented in A. A. Mills and R. Clift, "Reflections of the 'Burning Mirrors of Archimedes' with a Consideration of the Geometry and Intensity of Sunlight Reflected From Plane Mirrors," *European Journal of Physics,* Vol. 13, No. 6 (November 1992), p. 268–279.
- The passage from John Tzetzes is taken from Ivor Thomas's translation in the Loeb Classical Library (see bibliography).
- Another excellent discussion of Archimedes and his mirrors on which I relied is D. L. Simms, "Archimedes and the Burning Mirrors," *Technology and Culture,* Vol. 18, No. 1 (1975), pp. 1–24.
- The passage from Anthemius of Tralles is taken from Huxley. Adcock was helpful in understanding the power and range of Archimedes' mechanical catapults and other weapons.
- The medieval optical studies of Roger Bacon, Athanasius Kircher, and Abu Ali al-Hasan al-Haitham are recounted in W. E. Knowles Middleton's delightfully literate essay, "Archimedes, Kircher, Buffon and the Burning Mirrors," in *Isis,* Vol. 52 (1961), pp. 533–543.
- The discussion of Leonardo is adapted from Pedretti.

Chapter 3 The 4-Acre Solar Machine

- A. S. E. Ackermann's analysis of the Tacony sun plant was published in "The Shuman Sun-Heat Absorber," in *Nature,* April 4, 1912.
- Ackermann's elaborately detailed survey of solar power engineering up to 1914 is found in "The Utilization of Solar Energy," in the *Annual Report of the Board of Regents of The Smithsonian Institution* for the Year Ending June 30, 1915, pp. 141–166.
- Ackermann's death in 1951 inspired a dozen obituaries in British scientific journals that were helpful in piecing together the activities of his long life, including, especially, that which appeared in *The Engineer* of April 13, 1951. Ackermann also maintained a correspondence in later life with the secretary of the Smithsonian,

Charles D. Watson, in which he discusses his various scientific interests.

- Shuman's work in Tacony on low-pressure steam engines is discussed and analyzed in R. C. Carpenter, "Tests of a Simple Engine, Taking Steam at Less than Atmospheric Pressure," in the *Scientific American Supplement* of July 13, 1912.
- The Shuman 1000-horsepower demonstration engine is described at exhaustive length in an unsigned article (usually attributed to Shuman but probably authored by Ackermann), "Power from the Sun's Heat," in *Engineering News,* May 13, 1909, Vol. 61, No. 19.
- The prevailing optimism about solar power and other forms of alternative energy in the early twentieth century is illustrated in a number of contemporary articles in the scientific journals of the time, including "Harnessing Nature: Can the Free Energy of Space Be Utilized?," by Waldemar Kaempffert in *Scientific American,* April 5, 1913, and "The Commercial Utilization of Solar Radiation and Wind Power," in *Scientific American,* January 21, 1911.
- Sydney Moseley's gossipy memoir vividly depicts the situation in Egypt at the time of Lord Kitchener's return there in 1911, assesses Kitcherner's accomplishments (though in a sometimes hagiographic light), and paints a compelling personal portrait of the man.

Chapter 4 "A Substitute for Fuel in Tropical Countries"

- I am indebted to Ms. Kathy Davis at Butler Library, Columbia University, for making available to me a facsimile of John Ericcson's personal copy of William Adams's *Solar Heat.* Adams's memoir forms the basis for my discussion of his work in Bombay.
- Church's encylopedic two-volume biography of Ericsson details his wartime activities and devotes a chapter to his solar experiments. Ericsson's interest in solar steam power is also documented in his own extensive writings on the subject, including those in his book *Solar Investigations.*
- John I. Yellott's four-part "Captain John Ericsson: Pioneer in Solar Energy," in *The Sun At Work,* September and December, 1956; and March and June, 1957, focuses on his career as one of the first American solar engineers.

Chapter 5 Solar-Powered Irrigation in Egypt

- Edwards's classic *A Thousand Miles Up the Nile* and Balls's *Egypt of the Egyptians* were enormously helpful in enabling me to understand Egypt before and during Kitchener period, as, of course, was

Moseley's memoir. The Qom Ombo irrigation project is discussed at length in Balls, as well as Sir William Garstin's efforts to cut a canal through the Sudd, which is also detailed in Collins.

Chapter 6 Augustin Mouchot and the First "Sun Engine"

- For details of weather, geography, and cultural milieu surrounding the opening of the Paris Exposition, I am indebted to Horne's gripping and literate *The Fall of Paris.*
- One of Mouchot's many contemporary admirers and popularizers in France was the journalist Leon Simonin, who wrote passionately and frequently about solar power and Mouchot's inventions. See, for example, his 15-page "L'Emploi Industriel de La Chaleur Solaire," in *Revue des Deux Mondes,* May 1, 1876.
- Besides contemporary newspaper and journal accounts of Mouchot's exploits in France and Algeria, his own *La Chaleur Solaire* provides the scientific context in which his inventive genius took root.
- A treasure trove of Mouchot papers and reminiscences is to be found in the Centre Historique des Archives Nationale de France in two separate archives: F/17/2994: Missions scientifiques et littéraires. Mouchot, A., professeur de physique au lycée de Tours. Mission en Algérie pour continuer ses expériences sur la chaleur solaire, 1877, and LH 1947 No. 68: Dossier de Légion d'honneur de Mouchot Augustin Bernard, inventeur du système d'utilisation de la chaleur solaire comme force motrice.

Other useful articles about Mouchot, his inventions, and his times include the following:

- "Sur les Effects Mecaniques de l'Air Confine et Chauffe par les Rayons du Soleil," by Augustin Mouchot in *Comptes Rendus de l'Academie des Sciences,* Vol. 59 (1864), p. 527.
- "Resultats Obtenus dans les Essais d'Applications Industrielles de la Chaleur Solaire," by Augustin Mouchot in *Comptes Rendus de l'Academie des Sciences,* Vol. 81 (1875), pp. 571–574.
- "Resultats d'Experiences Faites en Divers Points de l'Algerie pour l'Emploi Industriel de la Chaleur Solaire," by Augustin Mouchot in *Comptes Rendus de l'Academie des Sciences,* Vol. 86 (1878), pp. 1019–1021.
- "Etude des Appareils Solaires," by A. Crova, in *Comptes Rendus de l'Academie des Sciences,* Vol. 94 (1882), pp. 943–945.
- "Nouveaux Resultats d'Utilization de la Chaleur Solaire Obtenus a Paris," by Abel Pifre in *Comptes Rendus de l'Academie des Sciences,* Vol. 91(1880), pp. 388–389.

- "A Solar Printing Press," in *Nature* (September 21, 1882), pp. 503–504.

Chapter 7 Egypt's Great Sun Machine

- Shuman's inauguration of the Maadi plant and his tour for Lord Kitchener on the banks of the Nile is captured nicely in Joseph A. Callanan's *The Great Sun Machine*, published in *The Lamp* (the Exxon/Mobil employee magazine) (spring 1975), pp. 4–9.
- A number of articles in contemporary scientific journals discuss, dissect, and evaluate the Maadi solar plant, including "An Egyptian Solar Power Plant, Putting the Sun to Work, *Scientific American* (January 25, 1913); "Sun Power: Its Commercial Utilisation," by George Hally, *Transactions of the Institution of Engineers and Shipbuilders in Scotland*, Vol. 57 (April 21, 1914), pp. 316–371; "Energy from the Sun," *Scientific American* (May 23, 1914); "The Sun Power Plant in Egypt," *Scientific American* (January 17, 1914); "The Most Rational Source of Power, Tapping the Sun's Radiant Energy Directly," *Scientific American* (November 1, 1913).
- Ackermann's financial cost-benefit analysis is found in "Sun Power Plant: A Comparative Estimate of the Cost of Power from Coal and Solar Radiation," *Scientific American Supplement*, Vol. 77, No. 1985 (January 17, 1914), p. 37.
- Frank Shuman had a remarkably thin skin when it came to criticism of his efforts in Egypt. His critics (including, for example, E. J. D. Coxe, who disputed the efficiency of Shuman's solar devices in a letter to the editor of *Scientific American* titled "The Shuman Solar Power Plant," in the issue of March 28, 1914) often were taken to task by Shuman for faulty engineering knowledge or errant physical analysis, as in "Sun-power Plants Not Visionary," a letter he wrote from Ackermann's consulting offices in England and published in Correspondence, *Scientific American* (June 27, 1914). He was also quick to write letters advertising any of the successes he had made there, as in "Feasibility of Utilizing Power from the Sun," in *Scientific American* (February 28, 1914).

Chapter 8 California Light and Solar Power

- The colorful history of the Bradbury Building is derived mainly from a short pamphlet available from the current owners of the building: "Bradbury, 1893, the Building's History, 304 South Broadway, Los Angeles, CA 90013." Late-night television addicts will recognize the light-filled interior from the innumerable films and TV movies made on location there. It continues to be one of Los Angeles's most interesting architectural spaces.

- Chandler's land and water exploits in Arizona are extensively documented in Zarbin and in Merrill. The Eneas machinery is described in Meinel.
- The California installations of the Eneas system are described at length in Charles F. Holder, "Solar Motors," *Scientific American* (March 16, 1901); and in "A Solar Motor," *The Railway and Engineering News* (February 23, 1901).
- The excitement felt about solar energy in the American West at the turn of the century is captured nicely in Robert H. Thurston "Utilising the Sun's Energy," *Cassier's Magazine* (August 1901), pp. 283–288.
- Eneas's Pasadena installation is discussed at length in Larkin and in "The Solar Motor at Pasadena, Cal.," by Alfred L. Davenport, in *Engineering News*, Vol. 45 (October, 1901), p. 330; "Harnessing Wind, Water, and Sun" by George B. Waldron in *Munsey's Magazine* (October, 1901), pp. 81–86; and "Harnessing the Sun," by F. B. Millard, in *World's Work*, Vol. 1 (April 1901).
- The Carl Spain interview is in the *Tempe Daily News*, May 5, 1976, p. 1-A.
- The Eneas patent for a solar generator discussed in this chapter is No. 670,916, issued March 26, 1901, to A. G. Eneas.
- Tellier's solar pumping system at his workshop outside Paris is described at length in "The Utilization of Solar Heat for the Elevation of Water," in *Scientific American* (October 3, 1885), as well as in his memoir about the frozen-food transport vessel *Frigorifique* (see bibliography). His essay on solar power in Africa appeared in the *Revue des Deux Mondes*.

Chapter 9 War and Petroleum: The End of an Era

- Winston Churchill's determination to commit the British Navy to petroleum is related in the prologue of Yergin's magisterial history of the oil business.
- Magnus, Mosley, and Warner all write at length about Kitchener's last tour in Egypt and his early service in the First World War.
- Wingate's trip to Balmoral to visit the king and secure Lloyd George's commitment to finance irrigation in Sudan is detailed in Sir Ronald Wingate's biography of his father (see bibliography).
- Frank Shuman's involvement in early submarine technology is detailed in U.S. Patent No. 1,310,253, issued posthumously to his

estate July 15, 1919; and in a June 18, 1939, interview with his widow published in the *Philadelphia Evening Bulletin* ("Navy Rebuffs Philadelphian Who Offered Navy Sub Idea"), in which she laments the U.S. Navy's lack of interest in his many submarine-related inventions. The Navy's concerns about German espionage in the United States in the years prior to the Great War, and its domestic intelligence operations, are documented in both volumes of Dorwart. Shuman's obituary also appeared in the *Philadelphia Evening Bulletin*.

- The story of modern photovoltaics is recounted in Perlin's recent *From Space to Earth*.

Bibliography

Adams, William, *Solar Heat: A Substitute for Fuel in Tropical Countries for Heating Steam Boilers, and Other Purposes*, Bombay: Education Society's Press, 1878.

Adcock, F. E., *The Greek and Macedonian Art of War*, Berkeley: University of California Press, 1957.

Armytage, W. H. G., *A Social History of Engineering*, Cambridge, MA: MIT Press, 1961.

Balls, W. Lawrence, *Egypt of the Egyptians*, New York: Charles Scribner's Sons, 1916.

Behrman, Daniel, *Solar Energy, The Awakening Science*, Boston: Little Brown, 1976.

Bernal, J. D., *Science and Industry in the Nineteenth Century*, Bloomington: Indiana University Press, 1970.

Branley, Franklyn M., *Solar Energy*, New York: Thomas Y. Crowell, 1957.

Braudel, Fernand, *Civilization and Capitalism, 15th–18th Century*, Vol. 3, *The Perspective of the World*, Berkeley: University of California Press, 1992.

Braun, Ernst, *Wayward Technology*, Westport, CT: Greenwood Press, 1984.

Brinkworth, B. J., *Solar Energy for Man*, New York: John Wiley & Sons, 1972.

Brinnin, John Malcolm, *Beau Voyage, Life Aboard the Last Great Ships*, New York: Dorset Press, 1981.

Buchanan, R. A., *The Power of the Machine, The Impact of Technology from 1700 to the Present*, London: Viking, 1992.

Burke, James, and Robert Ornstein, *The Axemakers Gift*, New York: Putnam, 1997.

Burke, John G., *The New Technology and Human Values*, Belmont, CA: Wadsworth, 1967.

Butler, Daniel Allen, *"Unsinkable" The Full Story, RMS Titanic,* Mechanicsburg, PA: Stackpole Books, 1998.

Butti, Ken, and John Perlin, *A Golden Thread: 2500 Years of Solar Architecture and Technology,* Palo Alto: Cheshire Books, 1980.

Carr, Donald E., *Energy and the Earth Machine,* New York: W. W. Norton, 1976.

Cheremisinoff, Paul N., and Thomas C. Regino, *Principles and Applications of Solar Energy,* Ann Arbor, MI: Ann Arbor Science Publishers, 1978.

Church, William Conant, *The Life of John Ericsson,* New York: Charles Scribner's Sons, 1911 (2 volumes).

Cohen, I. Bernard, *Revolution in Science,* Cambridge, MA: Harvard University Press, 1985.

Collins, Robert O., *The Waters of the Nile: Hydropolitics and the Jonglei Canal, 1900–1988,* Princeton, NJ: Markus Wiener Publishers, 1990.

Cooper, Gail, *Air-conditioning America,* Baltimore: Johns Hopkins University Press, 1998.

Daniels, Farrington, *The Challenge of Our Times,* Minneapolis: Burgess Publishing, 1953.

———, *Direct Use of the Sun's Energy,* New Haven, CT: Yale University Press, 1964.

Derby, George, ed., *The National Cyclopedia of American Biography,* Vol. 19, New York: James T. White & Company, 1926.

Dickinson, H. W., *A Short History of the Steam Engine,* London: Frank Cass Reprints, 1963.

Dorwart, Jeffrey M., *The Office of Naval Intelligence,* Annapolis: The Naval Institute Press, 1979.

Dorwart, Jeffrey M., and Jean K. Wolf, *The Philadelphia Navy Yard,* Philadelphia: University of Pennsylvania Press, 2001.

Edwards, Amelia B., *A Thousand Miles Up the Nile,* London: George Routledge and Sons, 1891.

Ericsson, John, *Solar Investigations,* New York: John Ross & Co., 1876.

Farber, Paul Lawrence, *Finding Order in Nature,* Baltimore: Johns Hopkins University Press, 2000.

Flinders Petrie, W. M., *Researches in Sinai,* London: John Murray, 1906.

Forbes, R. J., *Man the Maker,* New York: Henry Schuman, 1950.

Freitag, Alicia M., and Harry C. Silcox, *Historical Northeast Philadelphia: Stories and Memories,* Holland, PA: Brighton Press, 1994.

Gadler, Steve J., and Wendy W. Adamson, *Sun Power,* Minneapolis: Lerner Publications Company, 1980.

Gardiner, Patrick, *Theories of History,* Glencoe, IL: Free Press, 1959.

Gies, Frances, and Joseph Gies, *Cathedral, Forge, and Waterwheel,* New York: HarperCollins, 1994.

Halacy, D. S., Jr., *The Coming Age of Solar Energy,* New York: Harper & Row, 1963.

Halberstam, David, *The Reckoning,* New York: William Morrow, 1986.

Hardison, O. B., Jr., *Disappearing Through the Skylight,* New York: Viking, 1989.

Harman, Willis W., *An Incomplete Guide to the Future,* New York: W. W. Norton, 1979.

Hayes, Denis, *Rays of Hope: The Transition to the Post-petroleum World,* New York: W. W. Norton, 1977.

Higonnet, Patrice, ed., *Favorites of Fortune,* Cambridge, MA: Harvard University Press, 1991.

Hildebrandt, Stefan, and Anthony Tromba, *The Parsimonious Universe,* New York : Copernicus (Springer-Verlag), 1996.

Hills, Richard L., *Power from Steam,* Cambridge, England: Cambridge University Press, 1989.

Hoehling, A. A., ed., *They Fought Under the Sea,* Washington, D.C.: Army Times Publishing Company, 1962.

Horne, Alistair, *The Fall of Paris,* New York: Viking Penguin, 1987.

Hunter, Louis C., *A History of Industrial Power in the United States, 1780–1930,* Vol. 1: *Waterpower in the Century of the Steam Engine,* Charlottesville: University of Virginia Press, 1979.

———, *A History of Industrial Power in the United States, 1780–1930,* Vol. 2: *Steam Power,* Charlottesville: University of Virginia Press, 1985.

Hutchings, Elizabeth, and Edward Hutchings, eds., *Scientific Progress and Human Values,* New York: American Elsevier, 1967.

Huxley, G. L., *Anthemius of Tralles, A Study of Later Greek Geometry,* Cambridge, MA: Eaton Press, 1959.

Iatarola, Louis M., and Siobhan Gephart, *Tacony: Images of America,* Charleston: Arcadia, 2000.

Jansen, Ted J., *Solar Engineering Technology,* Englewood Cliffs, NJ: Prentice-Hall, 1985.

Jardine, Lisa, *Ingenious Pursuits: Building the Scientific Revolution,* New York: Doubleday, 1999.

Jennett, Sean, *The Loire,* New York: Hastings House, 1975.

Jordan, R. C., and W. E. Ibele, "Mechanical Energy from Solar Energy," *Proceedings of the World Symposium of Applied Solar Energy,* Menlo Park, CA: Stanford Research Institute, 1956.

Klemm, Friedrich, *A History of Western Technology,* Cambridge, MA: The MIT Press, 1964.

Knight, David C., *Harnessing the Sun,* New York: William Morrow, 1976.

Kozloff, Arielle P., *Egypt's Dazzling Sun: Amenhotep III and His World,* Bloomington: Indiana University Press, 1992.

Landes, David S., *The Unbound Prometheus,* Cambridge, England: Cambridge University Press, 1969.

Larkin, Edgar L., *Radiant Energy and Its Analysis, Its Relation to Modern Astrophysics,* Los Angeles, Baumgardt Publishing, 1903.

Leiss, William, *The Domination of Nature,* New York: George Braziller, 1972.

Lienhard, John H., *The Engines of Our Ingenuity,* Oxford, England: Oxford University Press, 2000.

Lord Lloyd, *Egypt Since Cromer,* New York: AMS Press, 1970 (reprint of the 1933 edition).

Magnus, Sir Philip, *Kitchener, Portrait of an Imperialist,* New York: E. P. Dutton & Co., 1968.

Manchester, Harland, *Trail Blazers of Technology,* New York: Charles Scribner's Sons, 1962.

Mansfield, Peter, *The British in Egypt,* New York: Holt, Rinehart and Winston, 1971.

McDaniels, David K., *The Sun, Our Future Energy Source,* New York: John Wiley & Sons, 1984.

Meinel, A. B., and M. P. Meinel, *Applied Solar Energy: An Introduction,* Reading, MA: Addison-Wesley, 1976.

Merrill, W. Earl, *One Hundred Echoes from Mesa's Past,* Mesa, AZ: Privately printed by the Mesa Historical Museum, 1975.

Mitcham, Carl, *Thinking Through Technology,* Chicago: University of Chicago Press, 1994.

Mitcham, Carl, and Robert Mackey, eds., *Philosophy and Technology: Readings in the Philosophical Problems of Technology,* New York: Free Press, 1983.

Mohamed, Duse, *In the Land of the Pharaohs,* New York: Frank Cass & Co., 1968 (reprint of the 1911 Stanley Paul & Co. edition).

Moseley, Sydney A., *With Kitchener in Cairo,* London: Cassell & Company, 1917.

Mouchot, A., *La Chaleur Solaire,* Paris: Albert Blanchard, 1980 (reprint of 1879 edition).

Muller, Herbert J., *The Children of Frankenstein,* Bloomington: Indiana University Press, 1970.

———, *Freedom in the Modern World,* New York: Harper & Row, 1966.

Mumford, Lewis, *Interpretations and Forecasts: 1922–1972,* New York: Harcourt Brace, 1973.

———, *The Myth of the Machine,* New York: Harcourt Brace, 1966.

———, *Technics and Civilization,* New York: Harcourt Brace, 1934.

O'Connor, David B., and Eric H. Cline, *Amenhotep III,* Ann Arbor: University of Michigan Press, 1998.

O'Connor, Richard, *The Cactus Throne,* New York: G. P. Putnam's Sons, 1971.

Pacey, Arnold, *Technology in World Civilization,* Cambridge, MA: MIT Press, 1990.

Park, Robert, *Voodoo Science,* Oxford, England: Oxford University Press, 2000.

Pedretti, Carlo, *The Literary Works of Leonardo Da Vinci,* Berkeley: University of California Press, 1977 (two volumes).

Perlin, John, *From Space to Earth: the Story of Solar Electricity,* Ann Arbor, MI: Aatec Publications, 1999.

Philadelphia Museum of Art, *The Second Empire, Art in France Under Napoleon III,* Detroit: Wayne State University Press, 1978.

Pinkney, David H., *Napoleon III and the Rebuilding of Paris,* Princeton, NJ: Princeton University Press, 1958.

Plessis, Alain, *The Rise and Fall of the Second Empire, 1852–1871,* Cambridge, England: Cambridge University Press, 1979.

Pope, Charles Henry, *Solar Heat,* Boston: Pope (self-published), 1903.

Porch, Douglas, *The Conquest of the Sahara,* New York: Alfred A. Knopf, 1984.

Porta, John Baptista (Derek J. Price, ed.), *Natural Magick,* New York: Basic Books, 1957.

Porter, Sir George, *Solar Energy: A Royal Society Discussion,* London: Royal Society, 1980.

Pursell, Carroll, Jr., *The Machine in America,* Baltimore: Johns Hopkins University Press, 1995.

Pursell, Carroll W., Jr., ed., *Technology in America,* Cambridge, MA: MIT Press, 1989.

Rau, H., *Solar Energy,* New York: Macmillan, 1964.

Rogerson, Barnaby, *Traveller's History of North Africa,* New York: Interlink, 1998.

Rolt, L. T. C., *Thomas Newcomen: The Prehistory of the Steam Engine,* London: Macdonald, 1963.

Rosenrock, Howard, *Machines with a Purpose,* New York: Oxford University Press, 1990.

Sale, Kirkpatrick, *Rebels Against the Future,* New York: Addison-Wesley, 1995.

Sarton, George, *Sarton on the History of Science,* Cambridge, MA: Harvard University Press, 1962.

Sayigh, A. A. M., *Solar Energy Engineering,* New York: Academic Press, 1977.

Seward, Desmond, *Napoleon's Family,* New York: Viking, 1986.

Silcox, Harry C., *A Place to Live and Work,* University Park: Pennsylvania State University Press, 1994.

Silcox, Harry C., ed., *The History of Tacony, Holmesburg and Mayfair,* Philadelphia: Brighton Press, 1992.

Smith, Charles, "Revisiting Solar Power's Past," *Technology Review,* Cambridge: Massachusetts Institute of Technology, July 1995.

Strandh, Sigvard, *A History of the Machine* (Ann Henning, trans.), New York: A&W Publishers, 1979.

Teller, Edward, *Energy from Heaven and Earth,* San Francisco: W. H. Freeman, 1979.

Tellier, Charles, *Histoire d'une Invention Moderne, le Frigorifique,* Paris: Delagrave, 1910.

Thurston, Robert H., *A History of the Growth of the Steam Engine,* London: Kennikat Press, 1939.

Tzetzes, John, *The Book of Histories (Chiliades),* in *Greek Mathematical Works* (Ivor Thomas, trans.), The Loeb Classical Library, Cambridge, MA: Harvard University Press, 1941.

van der Vat, Dan, *Stealth at Sea,* Boston: Houghton Mifflin, 1995.

Walker, Charles R., *Technology, Industry, and Man,* New York: McGraw-Hill, 1968.

Warner, Philip, *Kitchener, The Man Behind the Legend,* New York: Atheneum, 1986.

Wertheim, Margaret, *Pythagoras' Trousers,* New York: Times Books, Random House, 1995.

White, Ruth, *Yankee from Sweden,* New York: Henry Holt, 1960.

Wingate, Sir Ronald, *Wingate of the Sudan: The Life and Times of General Sir Reginald Wingate, Maker of the Anglo-Egyptian Sudan,* London: John Murray, 1955.

Yergin, Daniel, *The Prize: The Epic Quest for Oil, Money, and Power,* New York: Simon & Schuster, 1992.

Zarbin, Earl A., *Two Sides of the River: Salt River Valley Canals, 1867–1902,* Phoenix: Salt River Project, 1997.

Zim, Herbert S., *Submarines,* New York: Harcourt Brace, 1942.

Zweibel, Ken, *Harnessing Solar Power,* New York: Plenum Press, 1990.

Illustration Credits

Chapter 1

The Frank Shuman portrait is courtesy of the Free Library of Philadelphia. The photographs of the Tacony solar steam installation appeared originally in Butti & Perlin, *A Golden Thread,* and are republished here with permission of Van Nostrand Rheinhold Company, New York.

Chapter 2

Portrait of Athanasius Kircher courtesy of the Vatican Library. The drawings of Leonardo and his in-ground mirror courtesy of the Royal Library at Windsor Castle. The lightweight German burning mirror, courtesy of the Bettmann Archive. The illustration from Al-Haitham's optical works courtesy of the Bibliotheque Nationale de France.

Chapter 3

The diagrams of the 1000-horsepower steam installation appeared in *Engineering News* in 1909. Photograph of Kitchener from the collection of the National Portrait Gallery, London.

Chapter 4

The William Adams drawings are from his book, *Solar Heat*. The John Ericsson portrait is from Church's biography of Ericsson. The Ericsson solar motor and pyrometer are from his own *Solar Investigations.*

Chapter 5

The photograph of the *shadoof* provided courtesy of the University of Maryland. The steel engraving of the *sakieh* and the map of the Nile are taken from Amelia Edwards's travel memoir *A Thousand Miles up the Nile* (see bibliography).

Chapter 6

The engravings of Mouchot devices appeared orginally in *Nature* and the *Revue des Deux Mondes.* The drawing of the Adams solar cooker is taken from his memoir.

Chapter 7

Photographs of Frank Shuman's Maadi plant in Egypt appeared originally in *A Golden Thread* and are republished here with permission of Van Nostrand Rheinhold Company, New York.

Chapter 8

The Aubrey Eneas solar machine in Arizona is from the collection of the Tempe Historical Museum, Tempe, Arizona, and is reproduced here with permission. The engraving of Charles Tellier's cast-iron hot boxes appeared originally in *Nature* (1885).

Acknowledgments

It is not possible for me adequately to express my thanks to my agent, Christy Fletcher, who is the sort of agent writers dream about but seldom find, ably assisted by Liza Bolitzer, Whitney Lee, and the other professionals at Carlisle & Company in Manhattan. Christy has been a joy to work with, an able business partner, a diplomatic negotiator on my behalf, and a reservoir of encouragement. Most of all, she believed in the concept of this book when it was only an idea set forth in a short proposal. But for her, *The Power of Light* would not have been written.

I am deeply grateful also to my wise and patient editor, Ruth Mills, who saw a coherent structure in the story from the moment she finished reading an early draft (when its coherence was doubtful even to me), and has never wavered in her enthusiasm or her generosity. She deserves great credit for whatever virtues this book may have, though she is blameless for its faults. She brought to the editing a broad background in publishing and marketing books, prudently trimmed most of the many tangential discussions I wanted to include, and made many valuable suggestions for strengthening the text.

I want to express my gratitude also to Philip Ruppel, vice president of McGraw-Hill, who took a personal interest in this project when it needed support and brought the full resources of the McGraw-Hill organization to bear on its timely completion.

A very special thanks also to Janice Race, senior editing supervisor at McGraw-Hill, for her patience, diligence, and competence in the preparation and assembly of the copyedited manuscript and galleys.

I was very fortunate to have one of my wisest and oldest family friends, the late George Witmer Allen, comment extensively and helpfully on the chapters set in Egypt and Algeria, both locales where he lived and worked for many years during and after the Second World War. Sadly, he died in February 2002, before he could see the narrative in its final form.

I would like to thank Judy King in Houston for her expert help in preparing the manuscript for submission, and for weaning me of my perverse (and usually incorrect) use of the semicolon.

Dr. Harry C. Silcox, who has published many books and articles on Frank Shuman and his hometown of Tacony, Pennsylvania, kindly read the first chapter of the manuscript in draft and offered valuable encouragement and helpful advice. My reliance on his published works in my own research is also reflected in the bibliography.

I am grateful to John Perlin, author (with Ken Butti) of *A Golden Thread*, perhaps the most comprehensive treatment of the history of solar technology from ancient times to the development of photovoltaics (a subject he later took up separately in his wonderful *From Space to Earth*), for his supportive telephone conversations from California and much helpful advice while I was researching and writing.

Finally, how can I thank my wife, Debra, sufficiently for putting up with me and the mountains of dust and paper which littered our house during the 2 years I spent immersed in this project?

* * * * *

I'm indebted beyond measure to professional librarians across the country and overseas for the help I've received from them in researching this book, especially Cynthia Mayo in the Business and Technology Division of the Dallas Public Library in Dallas for her tireless efforts to track things down for me, ably assisted by Tim Bullard and the entire interlibrary loan staff at the Dallas Central Library, and all the staff and volunteers at the Renner-Frankford branch in North Dallas who put up with my endless visits; all of the managers and staff at the

Eugene McDermott Library of the University of Texas at Dallas; Kathy Davis at Columbia University's Butler Library for providing me with my own facsimile copy of William Adams's *Solar Heat*; Sandra L. Marton, head of Interlibrary Loan at Collin County Community College; Tracy Robinson at the Smithsonian Institution Archives for help in tracking down the correspondence of A. S. E. Ackermann; Siobhan Gephart, director of the Tacony Historical Society in Philadelphia, for her help with matters relating to Frank Shuman's residence there; Walt Stock at the Free Library of Philadelphia for biographical material on Frank Shuman, including hard-to-find photographic materials; Hedley Sutton at the British Library, Oriental and India Office Collections, for his help in researching William Adams; Jean-François Chanal and Cyril Chazal at the *Bibliotheque Nationale de France*, and Claire Bechu and Thierry Pin at the *Centre Historique des Archives Nationales de France*, for their invaluable help in researching the life and times of Augustin Mouchot; Ronald S. Wilkinson, Science, Technology & Business Division at the Library of Congress; Margaret Jerrido, Archivist and Head of the Urban Archives, Paley Library, Temple University, Philadelphia; Dr. Amy A. Douglass and John H. Akers of the Tempe Historical Museum in Arizona; Kjell Lagerstrom at the John Ericsson Society in New York City; Mary K. B. Carter at the Pasadena Public Library; Terri McGuire at the South Pasadena Public Library; and Claire M. Shipman at Collin County Community College Library, Spring Creek Campus in Plano, Texas, for her untiring help and support.

Index

About the Author

Frank Kryza began his writing career in 1972 in Connecticut as a staff reporter for *The New Haven Journal-Courier*. Born in 1950 in the Dominican Republic, he is a 1972 graduate of Yale College and a 1982 graduate of the Yale School of Management. From 1982 to 1997 he worked for Atlantic Richfield Company in finance, planning, and external affairs, both in the United States and overseas. He currently lives in Dallas, Texas. This is his first book.